CRO S
KEY GUIDE

Australian Mammals

First published in 2008. Reprinted 2017

Jacana Books, an imprint of
Allen & Unwin
83 Alexander Street
Crows Nest NSW 2065
Australia
Phone: (61 2) 8425 0100
Fax: (61 2) 9906 2218
Email: info@allenandunwin.com
Web: www.allenandunwin.com

National Library of Australia
Cataloguing-in-Publication entry:

Cronin, Leonard.
Cronin's key guide to Australian mammals.

Includes index.
ISBN 978 1 74175 110 9 (pbk.).

1. Mammals - Australia - Identification. I. Title.

599.0994

A LEONARD CRONIN PRODUCTION
Text: Leonard Cronin
Illustrations: Marion Westmacott
Distribution Maps: Laurel Cohn
Design: Robert Taylor
Editor: Laurel Cohn

Printed in China by Everbest Printing Co., Ltd

10 9 8 7 6 5

CRONIN'S KEY GUIDE

Australian Mammals

Leonard Cronin

Illustrations by Marion Westmacott

JACANA BOOKS

ALLEN&UNWIN

Contents

Introduction

Australian mammals are generally shy and cryptic, hiding by day in burrows, tree hollows or among dense vegetation, and emerging at dusk to forage and hunt. These strategies have allowed many to thrive in the arid regions, while the continent's isolation has nurtured the evolution of some very unusual mammals. Among them are the egg-laying, duck-billed platypus, the marsupial mole that 'swims' through desert sands, and the magnificent red kangaroo. Separation from the other continents has also limited the variety of mammals found in Australia. Until the arrival of humans some 55,000 years ago, only five of the world's 22 major mammalian groups existed here, and of these the bats and rodents made up more than half of the total species, as they still do today. Yet nowhere else do the three types of mammal — the monotremes, marsupials and placental mammals — exist side-by-side.

While all mammals feed their young on milk produced by the female's mammary glands, the monotremes, marsupials and placental mammals differ in the development of their newborn. The monotremes, represented in Australia by the platypus and echidna, lay soft-shelled eggs and incubate them in a temporary pouch or in a nest at the end of a burrow. Marsupials give birth to tiny embryonic young, naked and blind, with bud-like hind limbs but well-developed forelimbs armed with sharp claws. The newborn uses its forelimbs to crawl up the mother's belly enabling it to clamp its mouth firmly onto a teat. The larger marsupials suckle their young in a well-developed pouch, while some of the smaller species lack a pouch and drag them around clinging to their teats. Placental mammals nurture their embryonic young in the womb, providing all their needs through a very efficient placenta. The newborn are relatively large with fully developed limbs.

This book guides readers through the different groups of Australian mammals and includes profiles of all the widespread and distinctive species together with a number of marine mammals. The species are arranged alphabetically within their families or subfamilies. The species accounts include information about where the animal lives, its lifestyle, reproduction, development, diet, and visible traces of its presence. I have attempted to describe the patterns of activity of each species, including sheltering, territoriality, social interactions and feeding behaviour, but only a small number of Australian mammals have been reasonably well studied and little is known about the lives of many species. Information about diet is often obtained from the examination of scats and can be quite variable, depending on the season and the local availability of particular food types. Many Australian animals either do not drink or are able to exist for long periods without drinking by extracting sufficient water from their food.

The distribution maps indicate areas where an animal may be found, although this depends on the presence of suitable habitat with sufficient food, water and shelter. The distribution of a species is usually quite discontinuous within its range, and the blank areas on the map showing where a species is not found are often more informative than the coloured areas.

The information about status is derived from the Federal, State and Territory lists of threatened species, together with the Red List of Threatened Species compiled by the International Union for Conservation of Nature and Natural Resources (IUCN). This paints a bleak picture, ranking Australia as second only to the United States in the number of threatened animal species. More than 20 per cent of Australia's mammals are threatened with extinction, and one third of the world's extinct mammals are Australian. Land clearing and its consequences, including the salinisation of rivers and landscapes, is the foremost threat to the majority of species on these lists. If we are even to begin halting the tide of extinctions, habitat destruction must be stopped and replaced with a massive program of revegetation.

Miniatures of all the species in the book are presented in the visual key on the following pages. This key includes a distribution map of the species, its head-body length and page number in the book.

MONOTREMES

Short-beaked Echidna 22

Platypus 23

CARNIVOROUS MARSUPIALS

Crest-tailed Mulgara 24

Northern Quoll 27

Dibbler 30

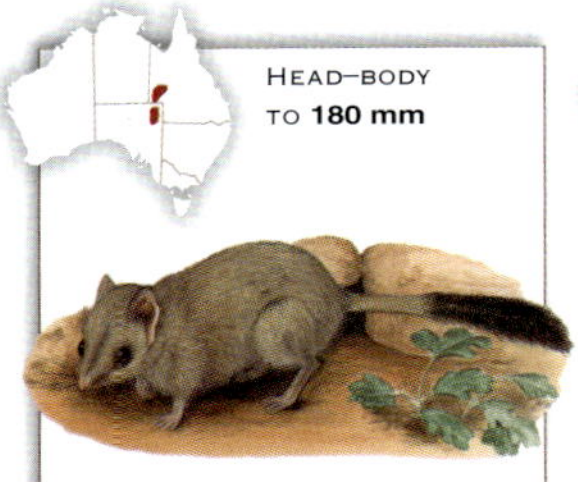

Kowari 25

Spotted-tailed Quoll 28

Sandstone Dibbler 31

Western Quoll 26

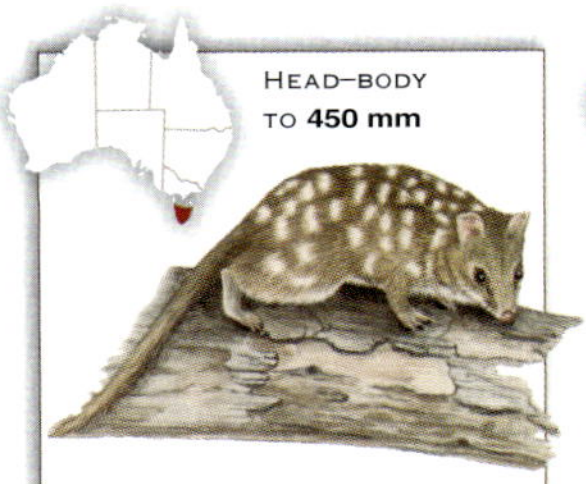

Eastern Quoll 29

Fat-tailed Pseudantechinus 32

CARNIVOROUS MARSUPIALS

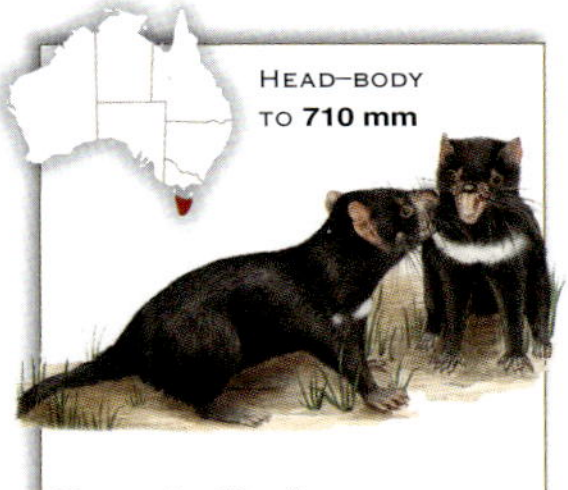
HEAD–BODY TO 710 mm

Tasmanian Devil **33**

HEAD–BODY TO 165 mm

Yellow-footed Antechinus **34**

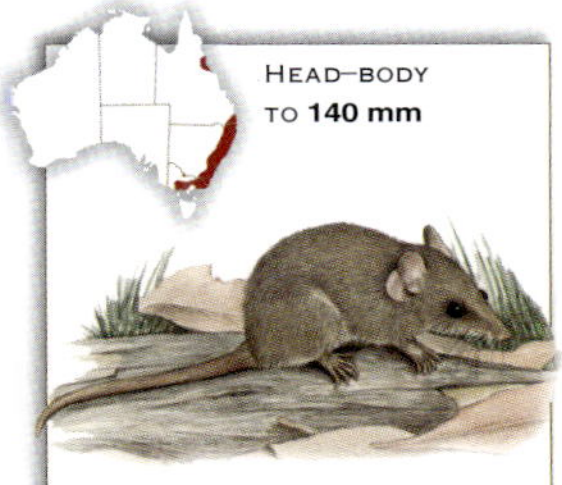
HEAD–BODY TO 140 mm

Brown Antechinus **35**

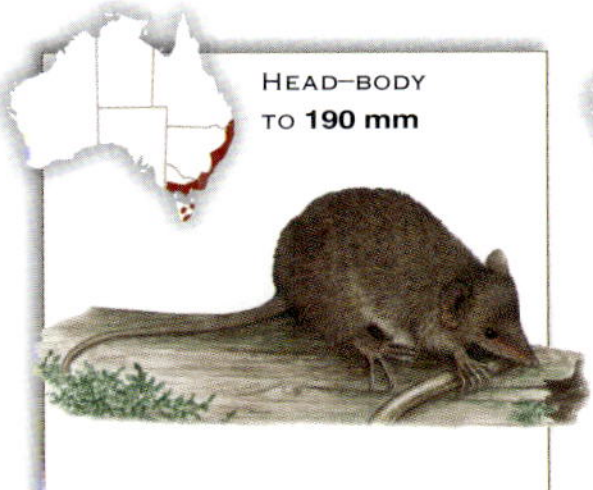
HEAD–BODY TO 190 mm

Dusky Antechinus **36**

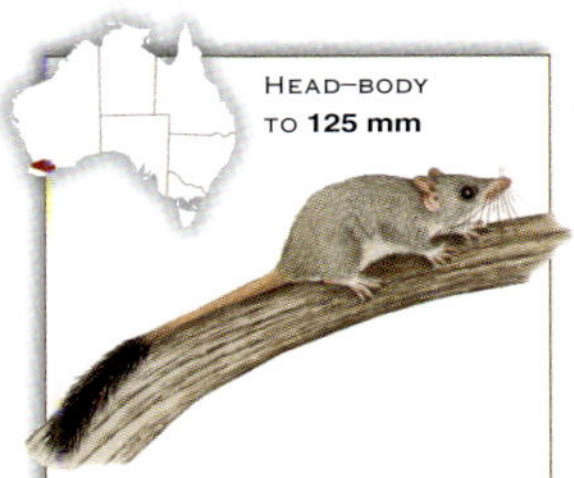
HEAD–BODY TO 125 mm

Red-tailed Phascogale **37**

HEAD–BODY TO 260 mm

Brush-tailed Phascogale **38**

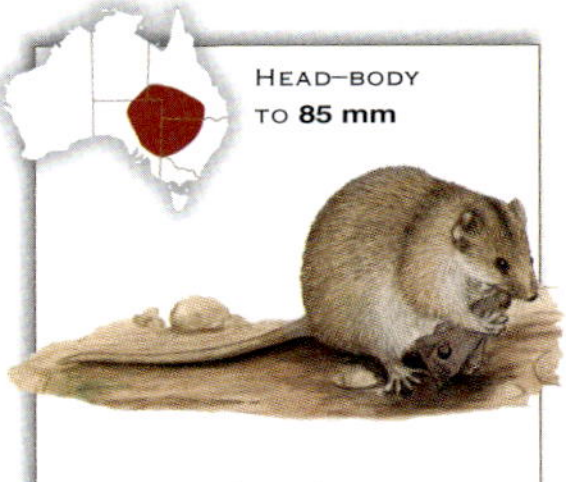
HEAD–BODY TO 85 mm

Paucident Planigale **39**

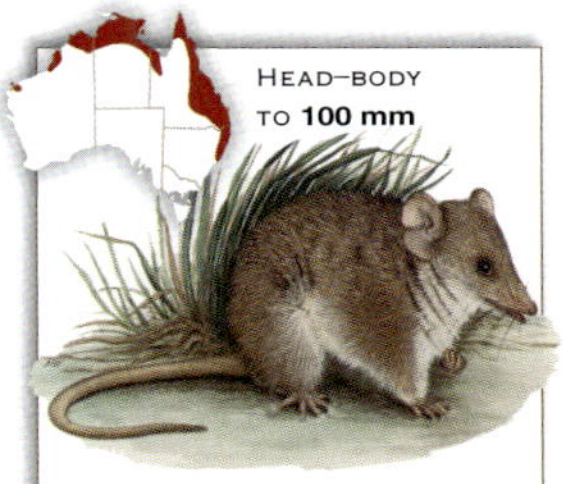
HEAD–BODY TO 100 mm

Common Planigale **40**

HEAD–BODY TO 75 mm

Narrow-nosed Planigale **41**

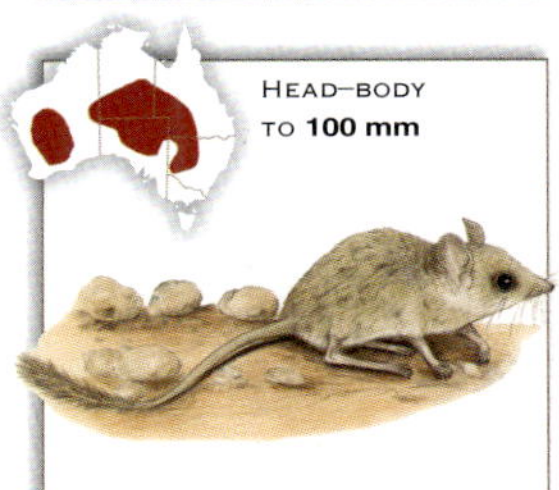
HEAD–BODY TO 100 mm

Kultarr **42**

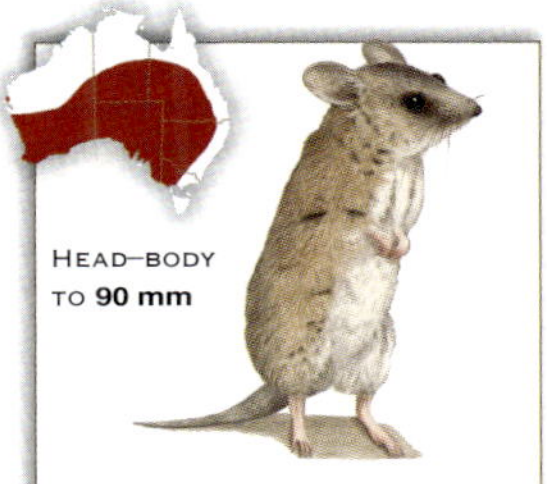
HEAD–BODY TO 90 mm

Fat-tailed Dunnart **43**

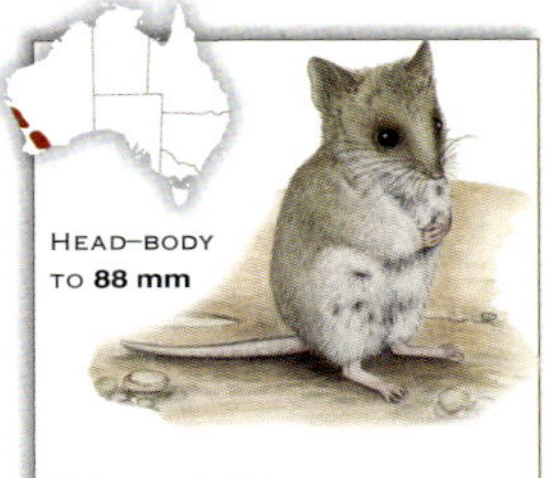
HEAD–BODY TO 88 mm

White-tailed Dunnart **44**

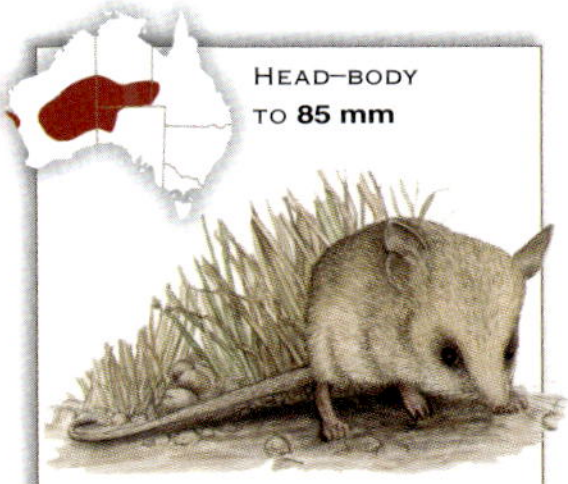
HEAD–BODY TO 85 mm

Hairy-footed Dunnart **45**

HEAD–BODY TO 120 mm

White-footed Dunnart **46**

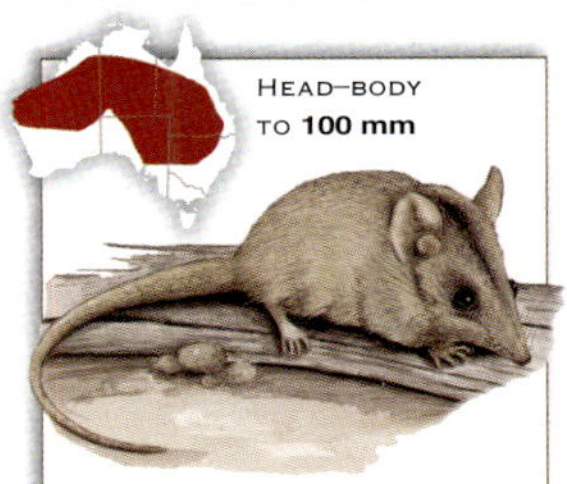
HEAD–BODY TO 100 mm

Stripe-faced Dunnart **47**

CARNIVOROUS MARSUPIALS

HEAD–BODY TO **105 mm**

Common Dunnart **48**

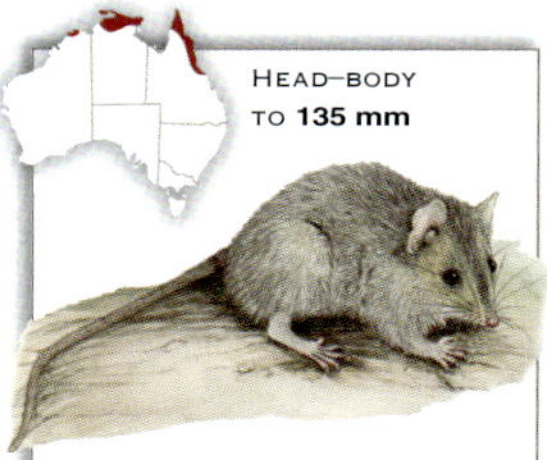
HEAD–BODY TO **135 mm**

Red-cheeked Dunnart **49**

HEAD–BODY TO **275 mm**

Numbat **50**

BANDICOOTS/BILBY

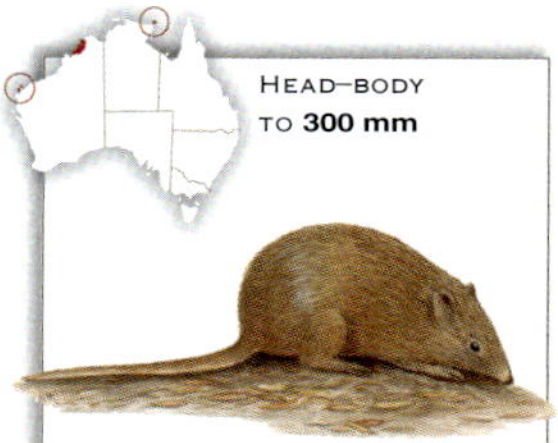
HEAD–BODY TO **300 mm**

Golden Bandicoot **51**

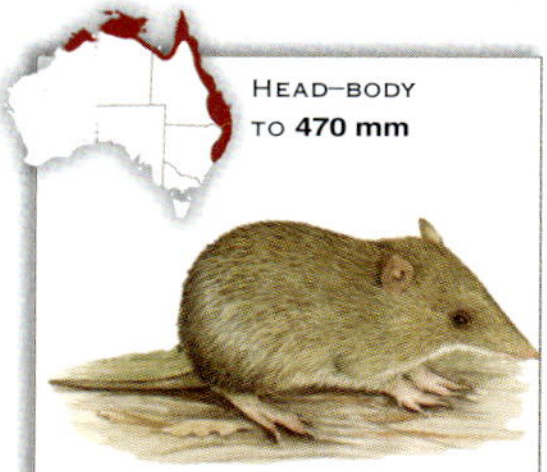
HEAD–BODY TO **470 mm**

Northern Brown Bandicoot **52**

HEAD–BODY TO **360 mm**

Southern Brown Bandicoot **53**

HEAD–BODY TO **236 mm**

Western Barred Bandicoot **54**

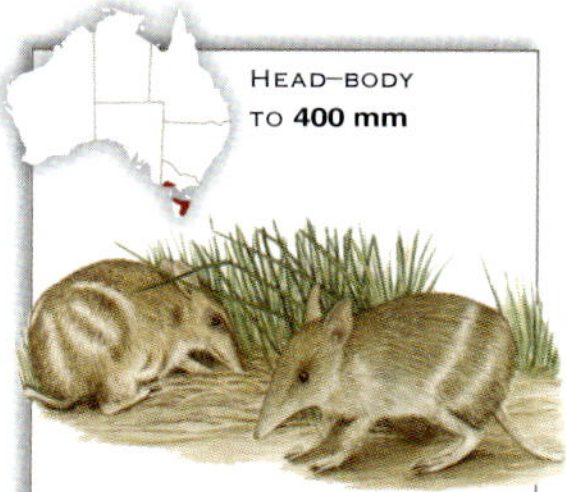
HEAD–BODY TO **400 mm**

Eastern Barred Bandicoot **55**

HEAD–BODY TO **430 mm**

Long-nosed Bandicoot **56**

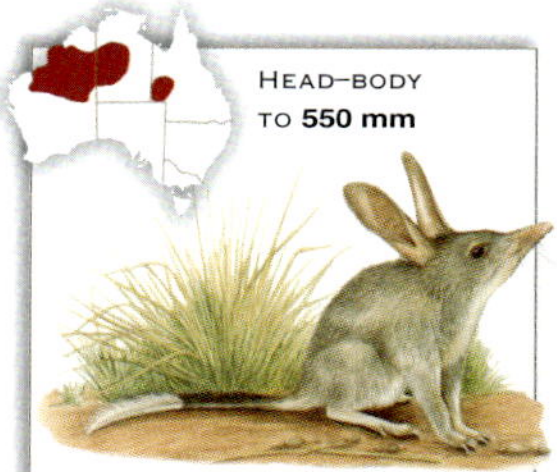
HEAD–BODY TO **550 mm**

Bilby **57**

HEAD–BODY TO **400 mm**

Rufous Spiny Bandicoot **58**

KOALA/WOMBATS

Head–body to 820 mm

Koala 59

Head–body to 940 mm

Southern Hairy-nosed Wombat 61

Head–body to 1100 mm

Northern Hairy-nosed Wombat 60

Head–body to 1200 mm

Common Wombat 62

POSSUMS AND GLIDERS

Head–body to 120 mm

Mountain Pygmy-possum 63

Head–body to 110 mm

Western Pygmy-possum 65

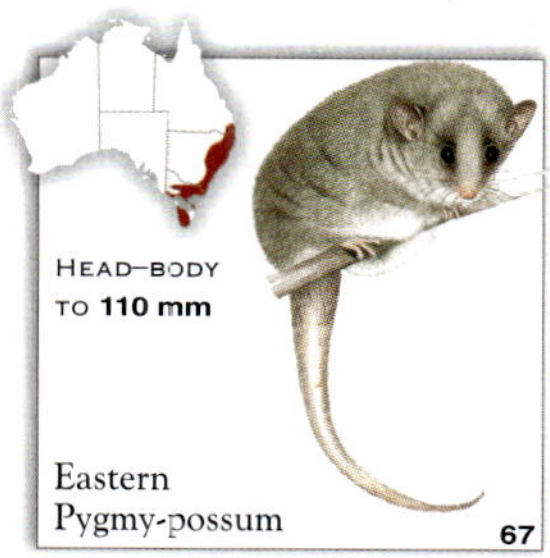

Head–body to 110 mm

Eastern Pygmy-possum 67

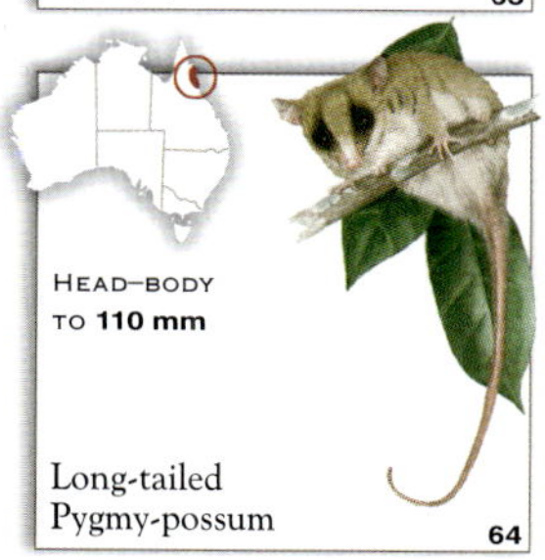

Head–body to 110 mm

Long-tailed Pygmy-possum 64

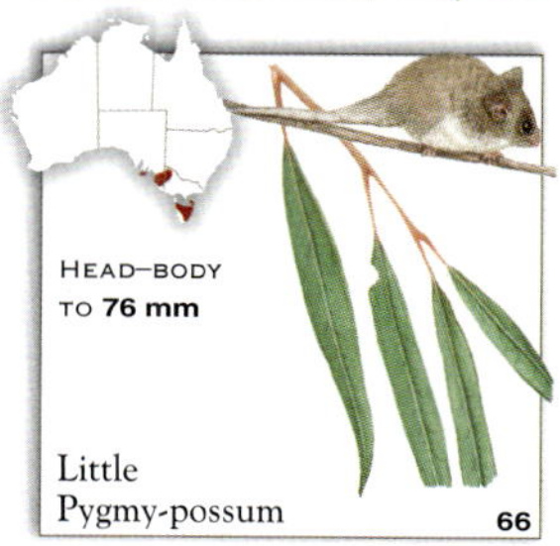

Head–body to 76 mm

Little Pygmy-possum 66

Head–body to 280 mm

Striped Possum 68

POSSUMS AND GLIDERS

HEAD–BODY TO 170 mm

Leadbeater's Possum 69

HEAD–BODY TO 450 mm

Greater Glider 74

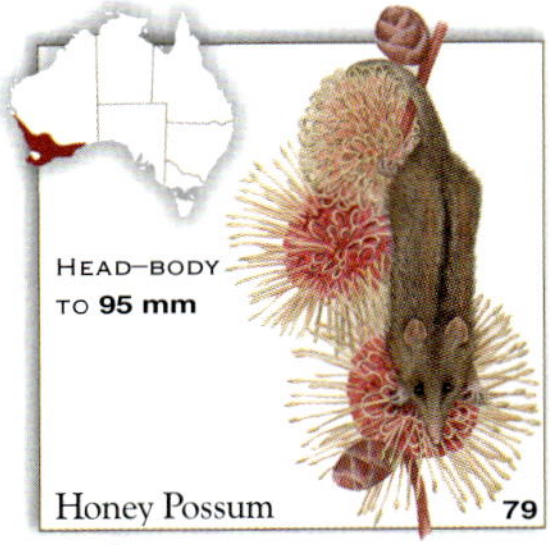
HEAD–BODY TO 95 mm

Honey Possum 79

HEAD–BODY TO 320 mm

Yellow-bellied Glider 70

HEAD–BODY TO 385 mm

Rock Ringtail Possum 75

HEAD–BODY TO 80 mm

Feathertail Glider 80

HEAD–BODY TO 210 mm

Sugar Glider 71

HEAD–BODY TO 380 mm

Green Ringtail Possum 76

HEAD–BODY TO 580 mm

Spotted Cuscus 81

HEAD–BODY TO 240 mm

Squirrel Glider 72

HEAD–BODY TO 400 mm

Herbert River Ringtail Possum 77

HEAD–BODY TO 400 mm

Southern Common Cuscus 82

HEAD–BODY TO 380 mm

Lemuroid Ringtail Possum 73

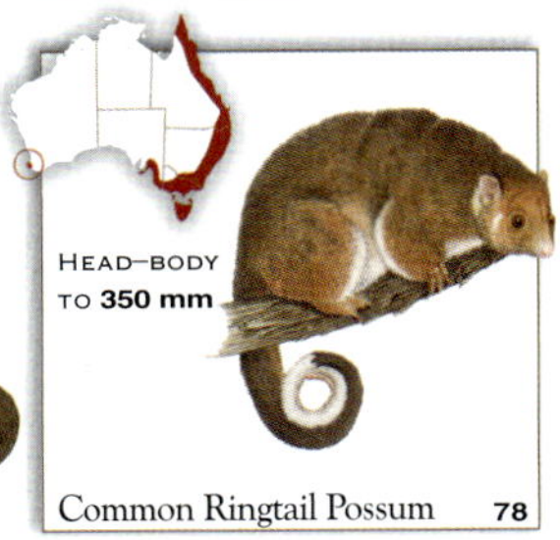
HEAD–BODY TO 350 mm

Common Ringtail Possum 78

HEAD–BODY TO 570 mm

Mountain Brushtail Possum 83

POSSUMS AND GLIDERS

HEAD–BODY TO **580 mm**

Common Brushtail Possum **84**

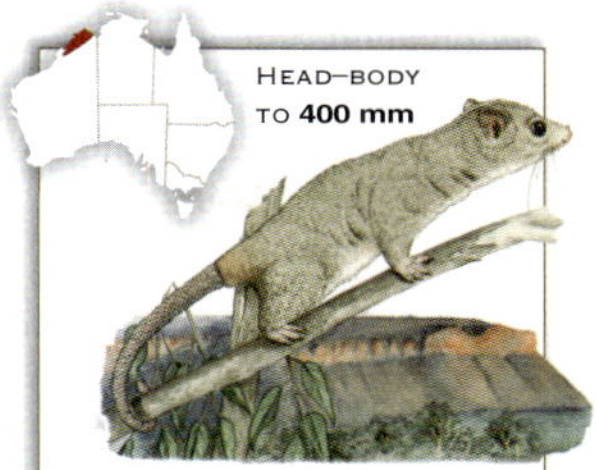

HEAD–BODY TO **400 mm**

Scaly-tailed Possum **85**

KANGAROOS

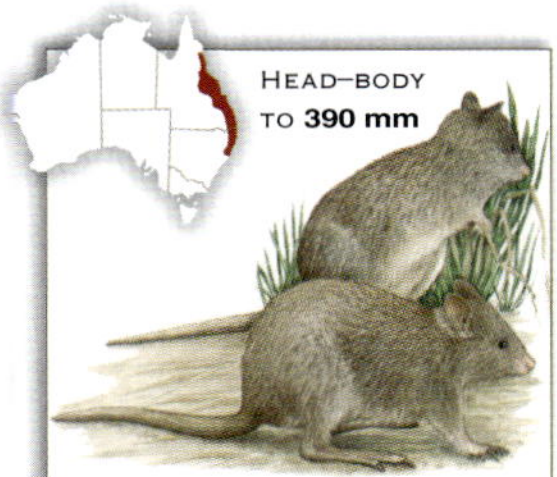

HEAD–BODY TO **390 mm**

Rufous Bettong **86**

HEAD–BODY TO **400 mm**

Burrowing Bettong **89**

HEAD–BODY TO **400 mm**

Long-nosed Potoroo **92**

HEAD–BODY TO **273 mm**

Musky Rat-kangaroo **87**

HEAD–BODY TO **380 mm**

Brush-tailed Bettong **90**

HEAD–BODY TO **750 mm**

Bennett's Tree-kangaroo **93**

HEAD–BODY TO **350 mm**

Tasmanian Bettong **88**

HEAD–BODY TO **415 mm**

Long-footed Potoroo **91**

HEAD–BODY TO **600 mm**

Lumholtz's Tree-kangaroo **94**

KANGAROOS

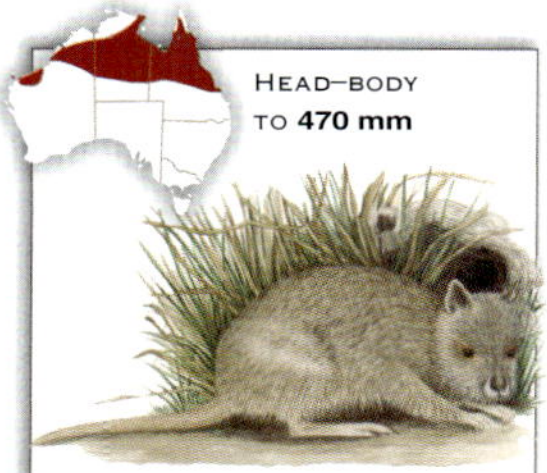

Spectacled Hare-wallaby **95**

Rufous Hare-wallaby **96**

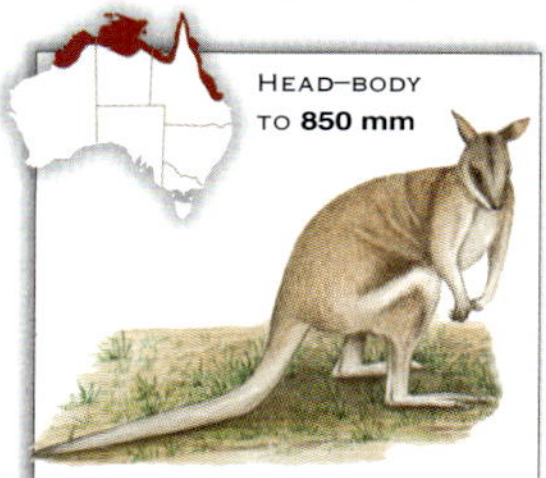

Agile Wallaby **97**

Antilopine Wallaroo **98**

Black Wallaroo **99**

Black-striped Wallaby **100**

Tammar Wallaby **101**

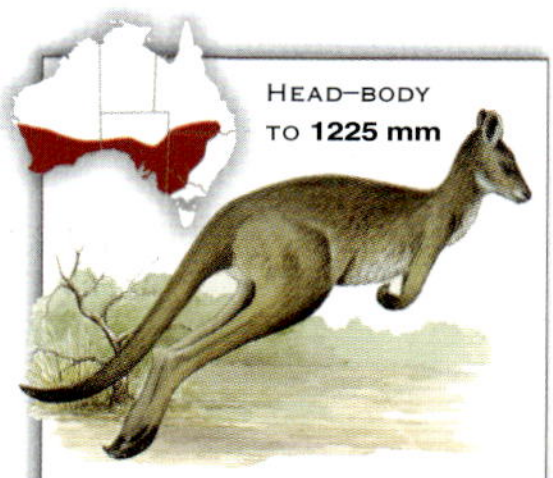

Western Grey Kangaroo **102**

Eastern Grey Kangaroo **103**

Western Brush Wallaby **104**

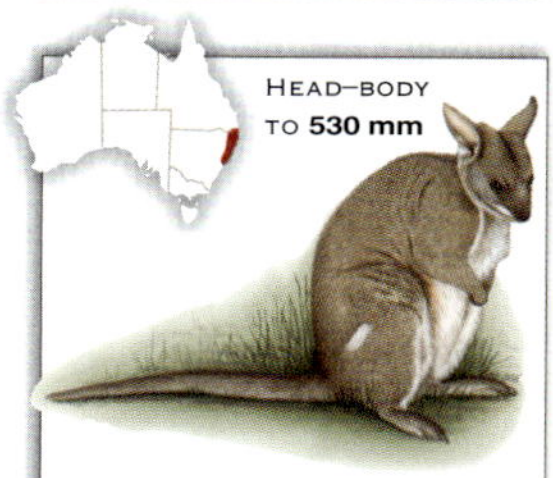

Parma Wallaby **105**

Whiptail Wallaby **106**

Common Wallaroo **107**

KANGAROOS

Head–body
to 925 mm

Red-necked Wallaby **108**

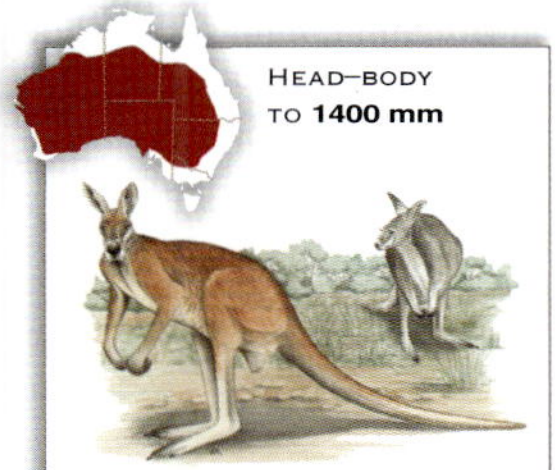

Head–body
to 1400 mm

Red Kangaroo **109**

Head–body
to 700 mm

Bridled Nailtail Wallaby **110**

Head–body
to 690 mm

Northern Nailtail Wallaby **111**

Head–body
to 550 mm

Short-eared Rock-wallaby **112**

Head–body
to 521 mm

Black-footed Rock-wallaby **113**

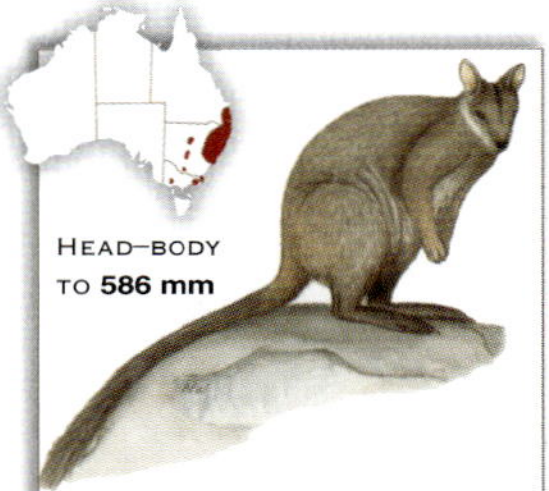

Head–body
to 586 mm

Brush-tailed Rock-wallaby **114**

Head–body
to 650 mm

Yellow-footed Rock-wallaby **115**

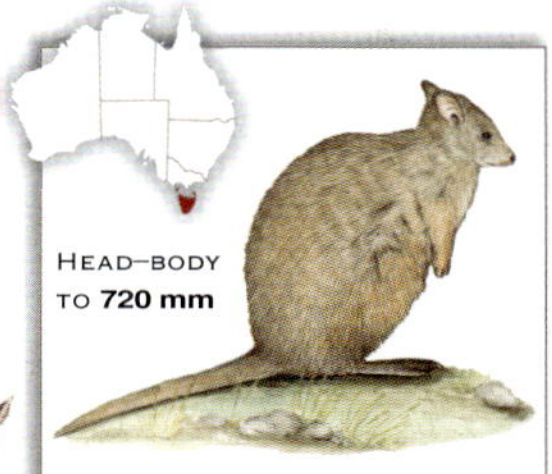

Head–body
to 720 mm

Tasmanian Pademelon **116**

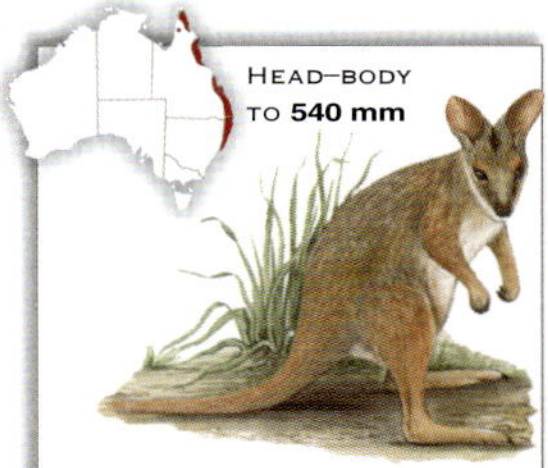

Head–body
to 540 mm

Red-legged Pademelon **117**

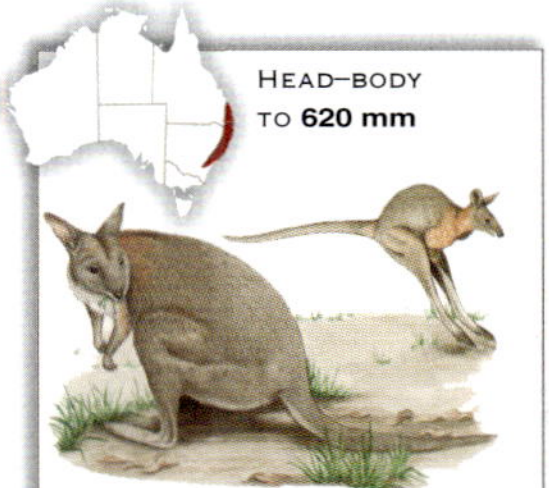

Head–body
to 620 mm

Red-necked Pademelon **118**

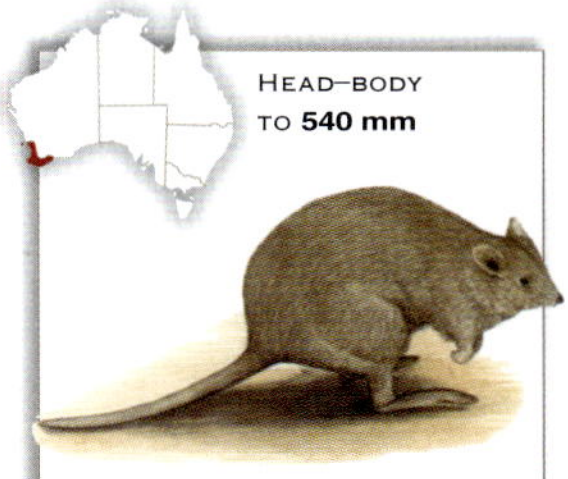

Head–body
to 540 mm

Quokka **119**

Head–body
to 847 mm

Swamp Wallaby **120**

MARSUPIAL MOLE

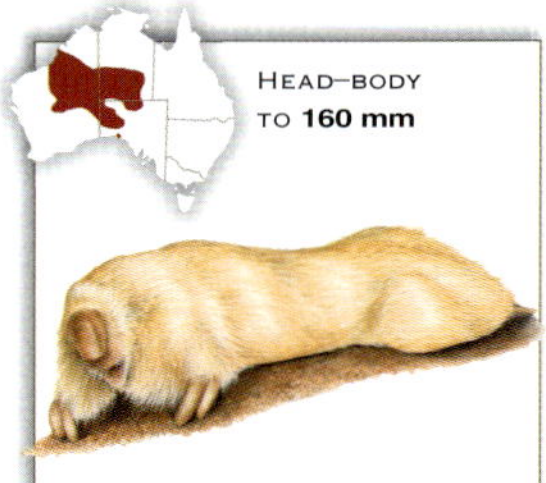
HEAD–BODY TO **160 mm**

Southern Marsupial Mole **121**

BATS

HEAD–BODY TO **67 mm**

Northern Blossom-bat **122**

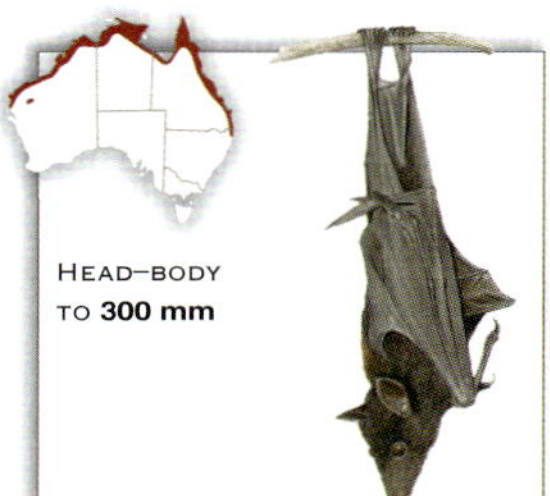
HEAD–BODY TO **300 mm**

Black Flying-fox **125**

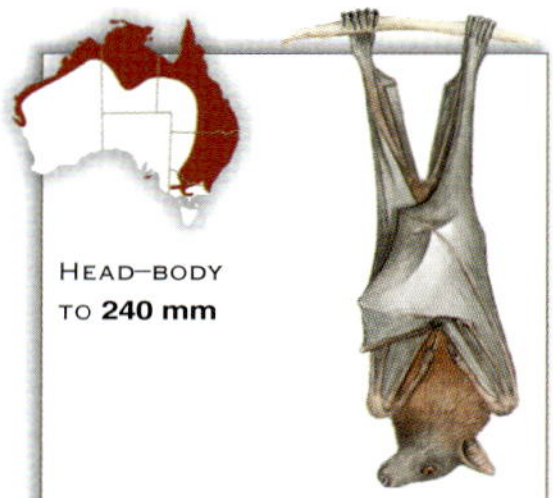
HEAD–BODY TO **240 mm**

Little Red Flying-fox **128**

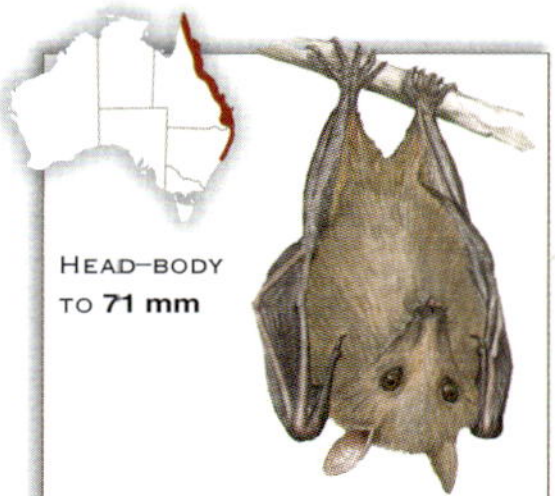
HEAD–BODY TO **71 mm**

Common Blossom-bat **123**

HEAD–BODY TO **240 mm**

Spectacled Flying-fox **126**

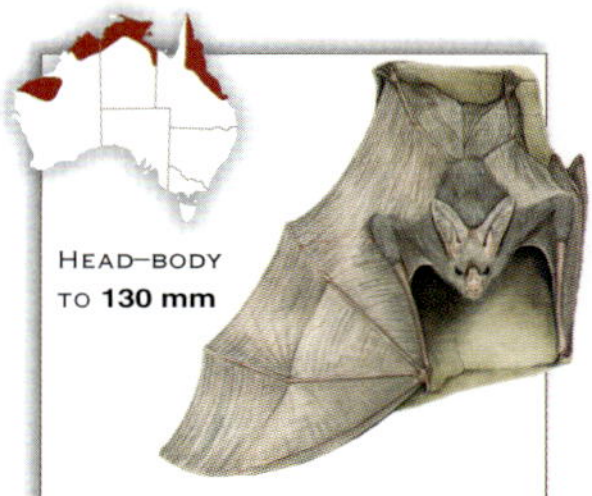
HEAD–BODY TO **130 mm**

Ghost Bat **129**

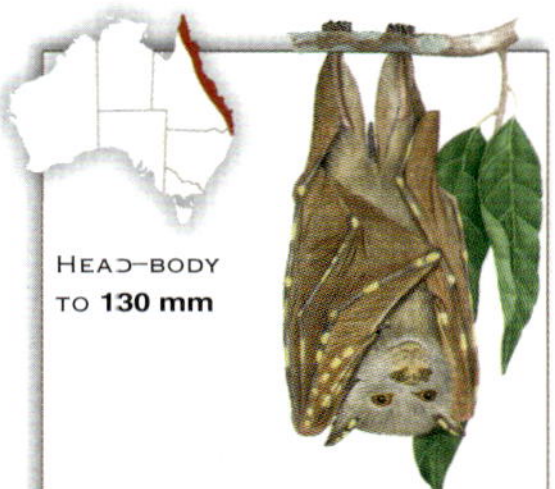
HEAD–BODY TO **130 mm**

Eastern Tube-nosed Bat **124**

HEAD–BODY TO **290 mm**

Grey-headed Flying-fox **127**

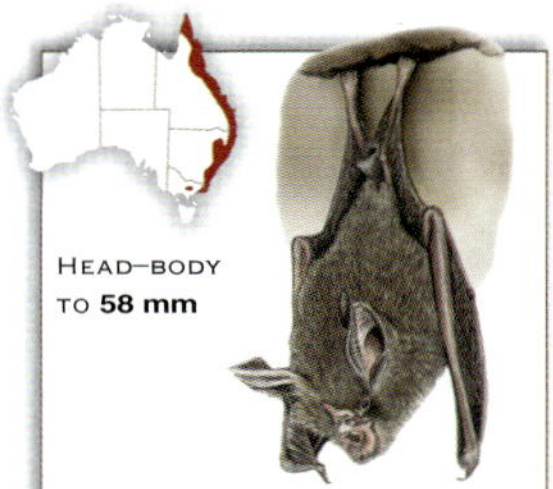
HEAD–BODY TO **58 mm**

Eastern Horseshoe Bat **130**

BATS

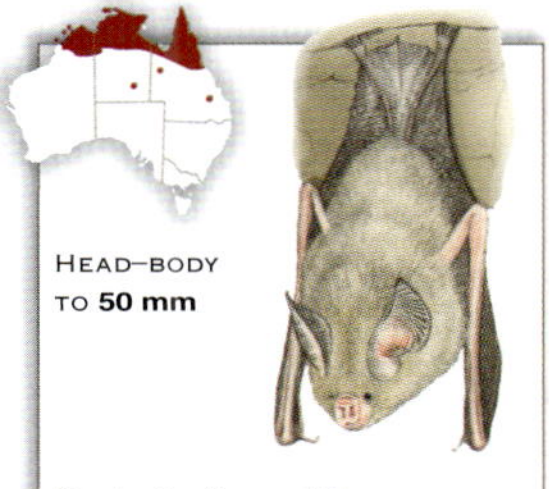
Head–body to 50 mm

Dusky Leaf-nosed Bat **131**

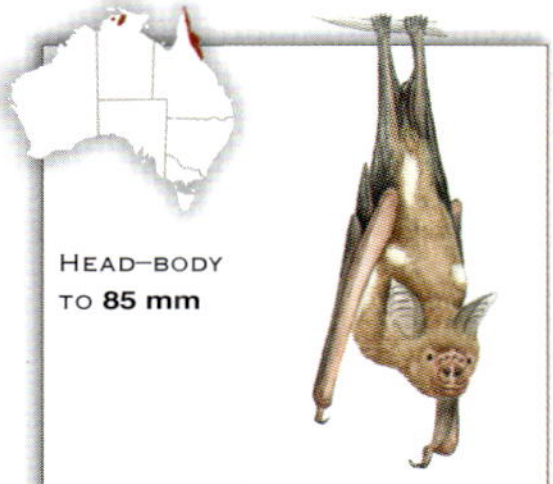
Head–body to 85 mm

Diadem Leaf-nosed Bat **132**

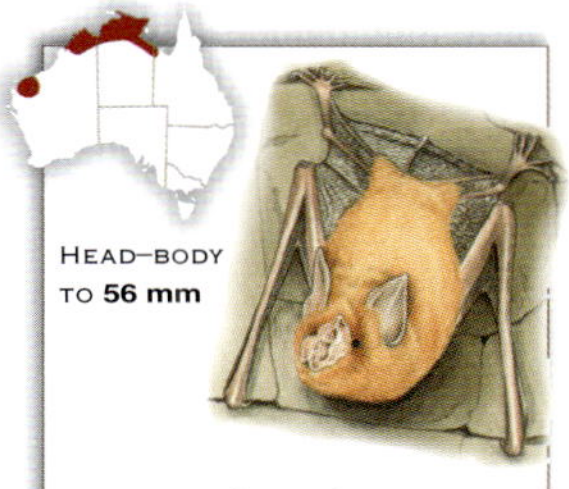
Head–body to 56 mm

Orange Leaf-nosed Bat **133**

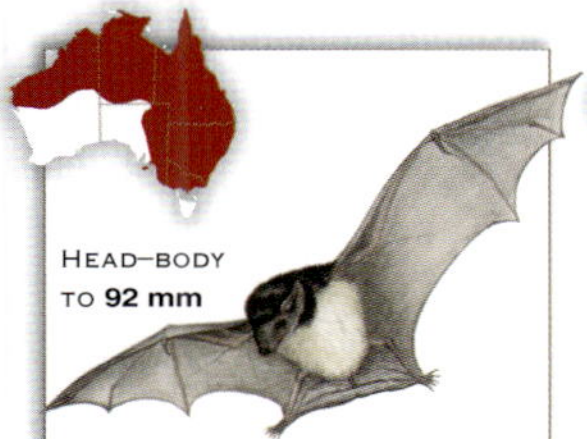
Head–body to 92 mm

Yellow-bellied Sheathtail-bat **134**

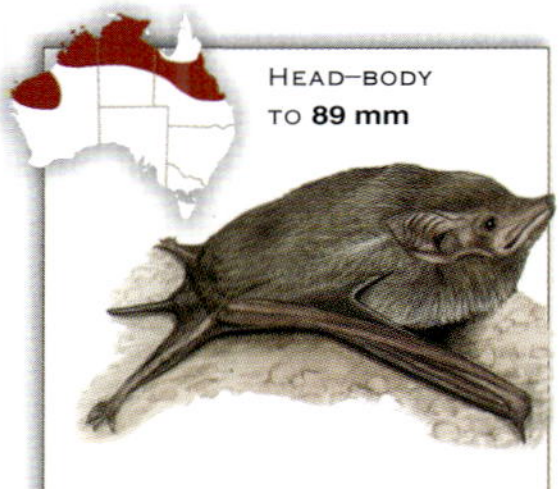
Head–body to 89 mm

Common Sheathtail-bat **135**

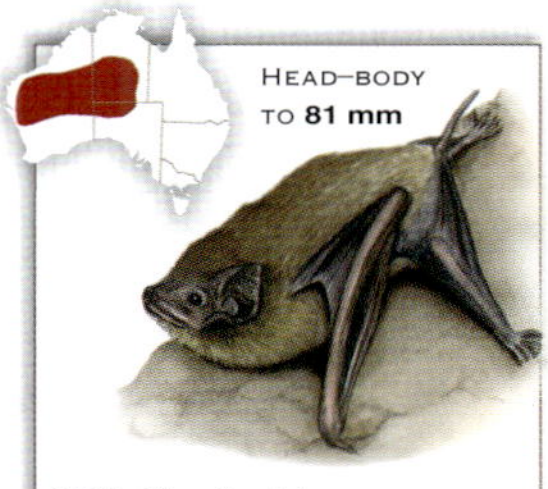
Head–body to 81 mm

Hill's Sheathtail-bat **136**

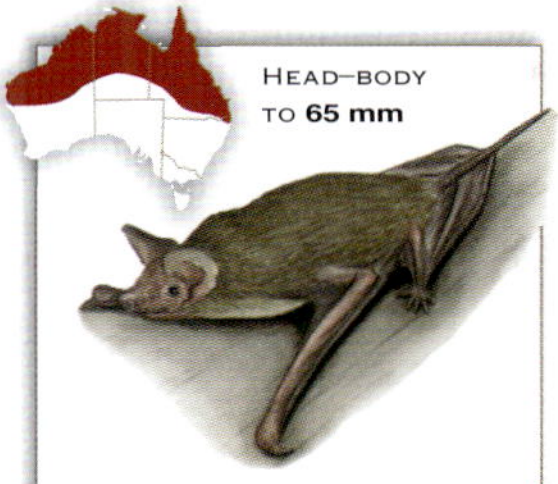
Head–body to 65 mm

Beccari's Freetail-bat **137**

Head–body to 55 mm

Little Northern Freetail-bat **138**

Head–body to 55 mm

Eastern Freetail-bat **139**

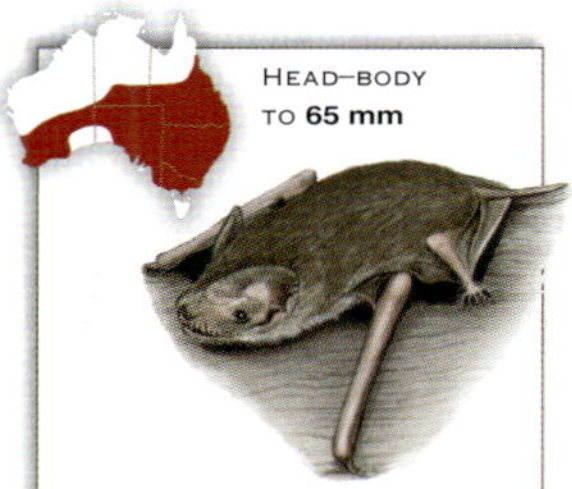
Head–body to 65 mm

Southern Freetail-bat **140**

Head–body to 100 mm

White-striped Freetail-bat **141**

Head–body to 48 mm

Little Bentwing Bat **142**

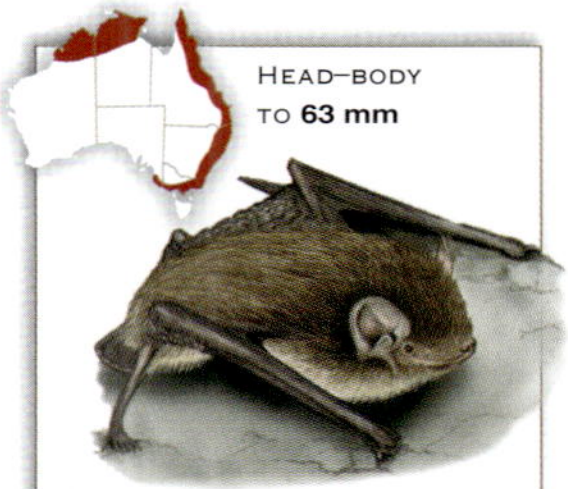
Head–body to 63 mm

Common Bentwing Bat **143**

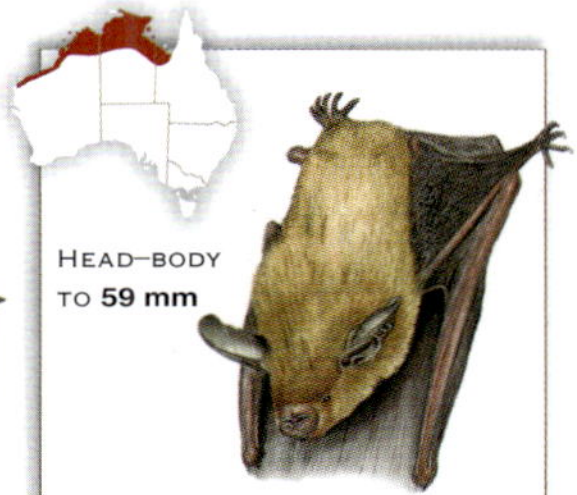
Head–body to 59 mm

Arnhem Long-eared Bat **144**

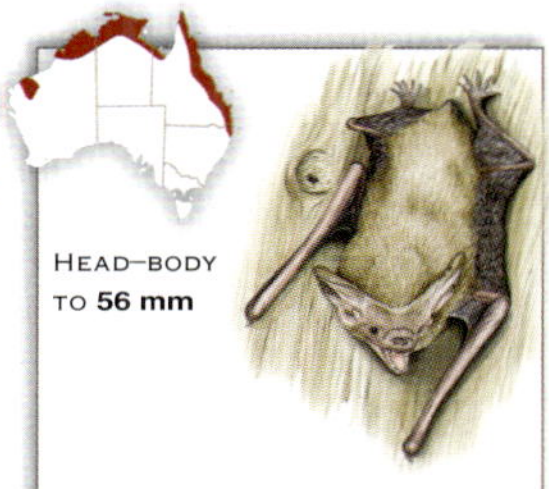
Head–body to 56 mm

Northern Long-eared Bat **145**

BATS

HEAD–BODY TO **50 mm**

Lesser Long-eared Bat **146**

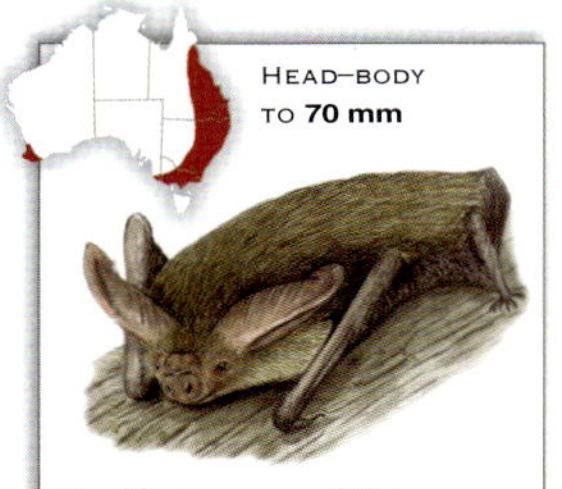

HEAD–BODY TO **70 mm**

Gould's Long-eared Bat **147**

HEAD–BODY TO **55 mm**

Large-eared Pied Bat **148**

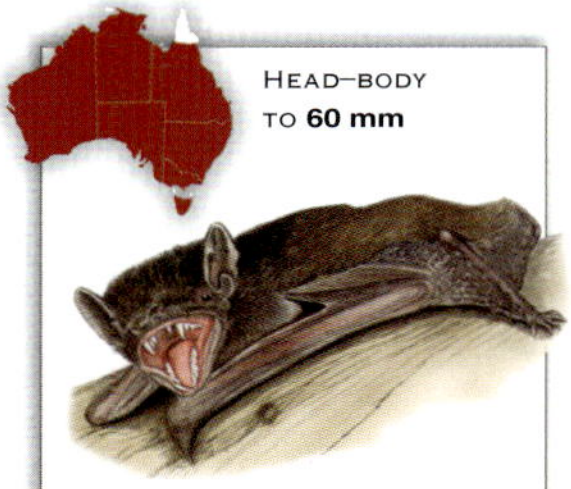

HEAD–BODY TO **60 mm**

Gould's Wattled Bat **149**

HEAD–BODY TO **61 mm**

Chocolate Wattled Bat **150**

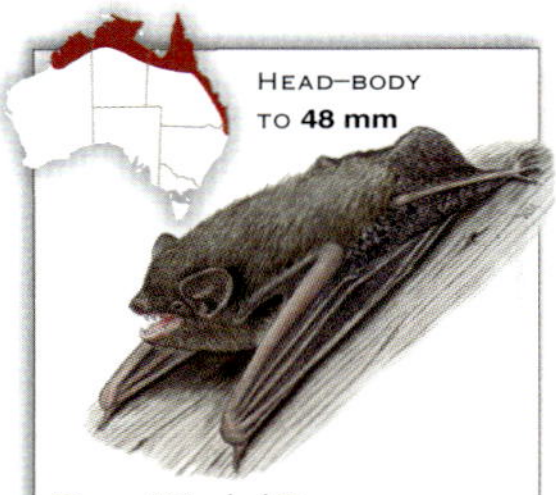

HEAD–BODY TO **48 mm**

Hoary Wattled Bat **151**

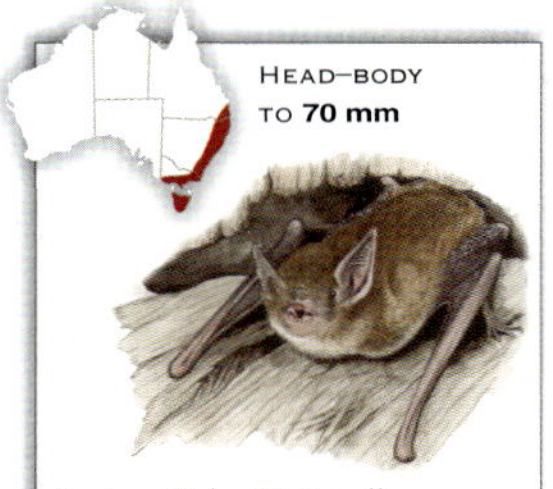

HEAD–BODY TO **70 mm**

Eastern False Pipistrelle **152**

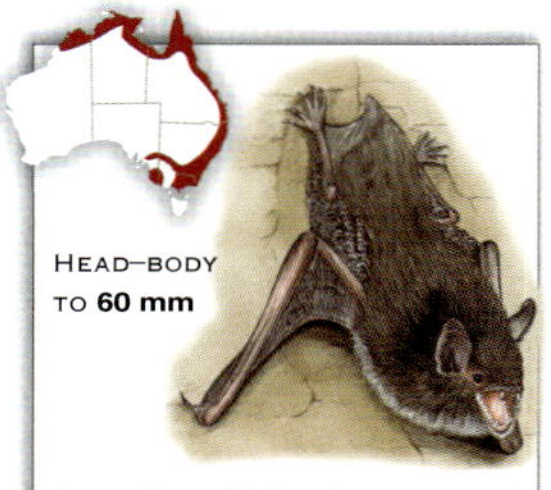

HEAD–BODY TO **60 mm**

Large-footed Myotis **153**

HEAD–BODY TO **43 mm**

Northern Pipistrelle **154**

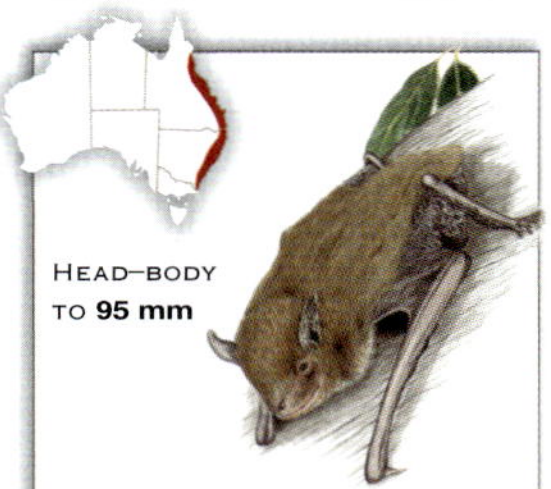

HEAD–BODY TO **95 mm**

Greater Broad-nosed Bat **155**

HEAD–BODY TO **60 mm**

Inland Broad-nosed Bat **156**

HEAD–BODY TO **55 mm**

Little Broad-nosed Bat **157**

HEAD–BODY TO **52 mm**

Northern Broad-nosed Bat **158**

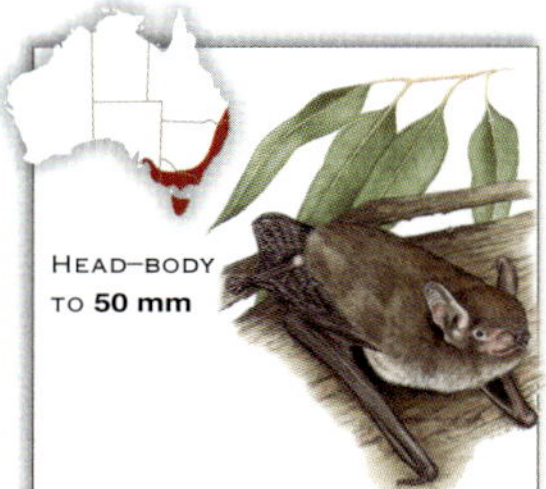

HEAD–BODY TO **50 mm**

Large Forest Bat **159**

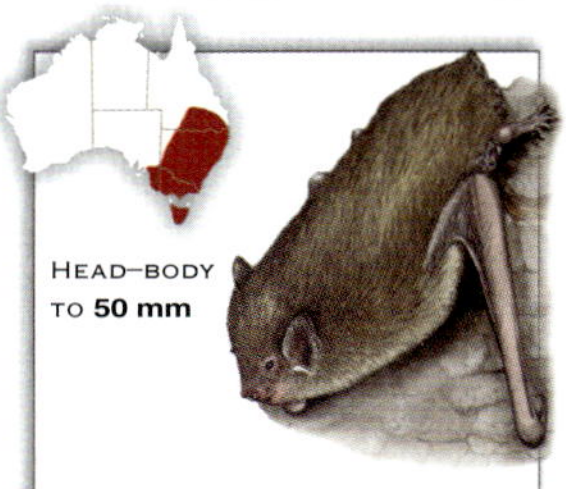

HEAD–BODY TO **50 mm**

Little Forest Bat **160**

RODENTS

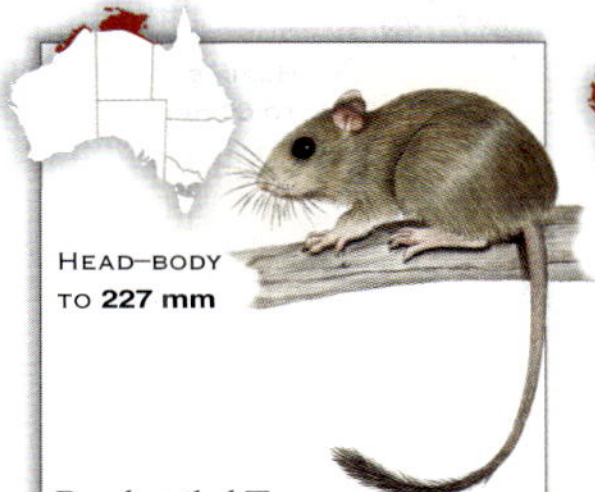

HEAD–BODY TO **227 mm**

Brush-tailed Tree-rat **161**

HEAD–BODY TO **115 mm**

Spinifex Hopping-mouse **166**

HEAD–BODY TO **75 mm**

Delicate Mouse **171**

HEAD–BODY TO **105 mm**

Forrest's Mouse **162**

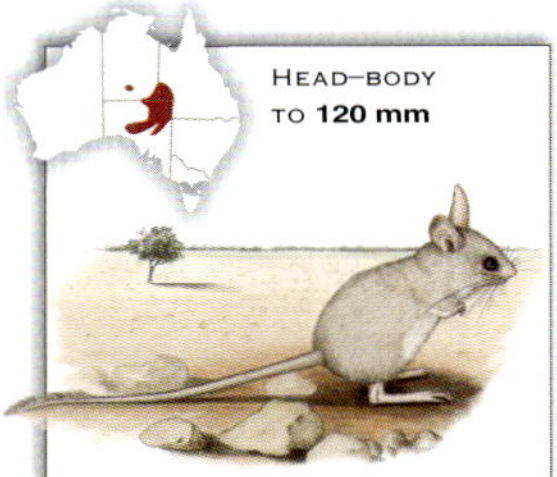

HEAD–BODY TO **120 mm**

Fawn Hopping-mouse **167**

HEAD–BODY TO **85 mm**

Sandy Inland Mouse **172**

HEAD–BODY TO **195 mm**

Broad-toothed Rat **163**

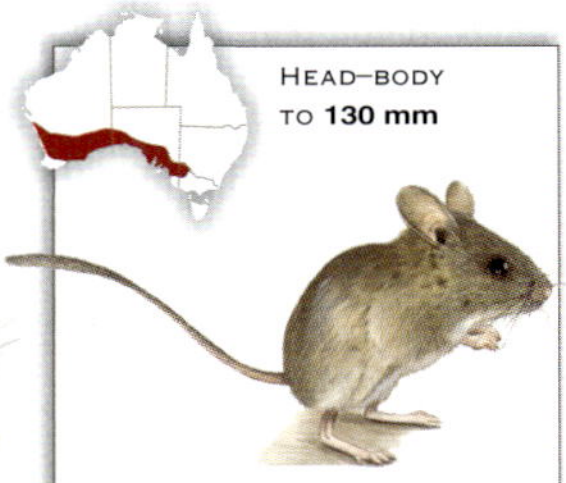

HEAD–BODY TO **130 mm**

Mitchell's Hopping-mouse **168**

HEAD–BODY TO **150 mm**

Long-tailed Mouse **173**

HEAD–BODY TO **310 mm**

Black-footed Tree-rat **164**

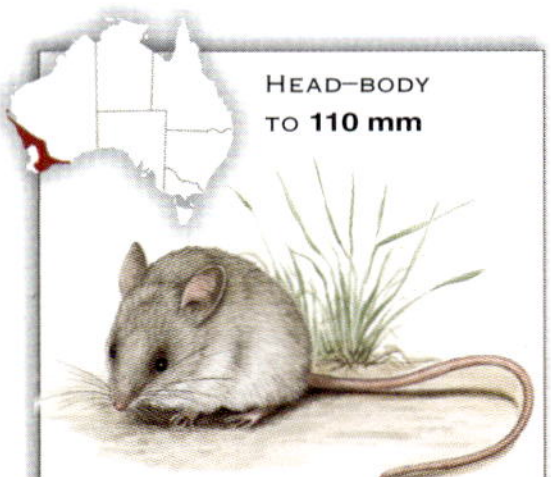

HEAD–BODY TO **110 mm**

Ash-grey Mouse **169**

HEAD–BODY TO **140 mm**

Western Chestut Mouse **174**

HEAD–BODY TO **245 mm**

Golden-backed Tree-rat **165**

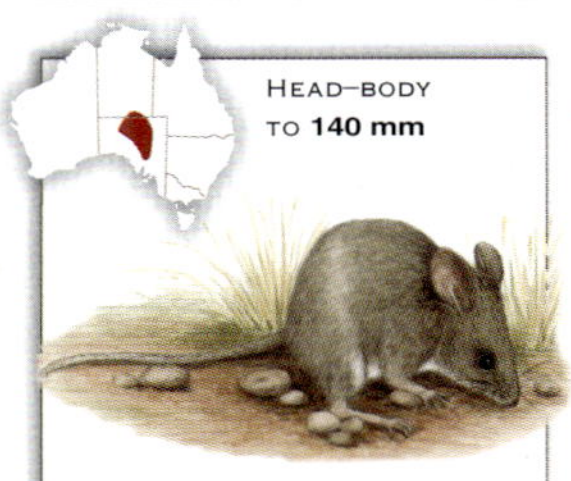

HEAD–BODY TO **140 mm**

Plains Rat **170**

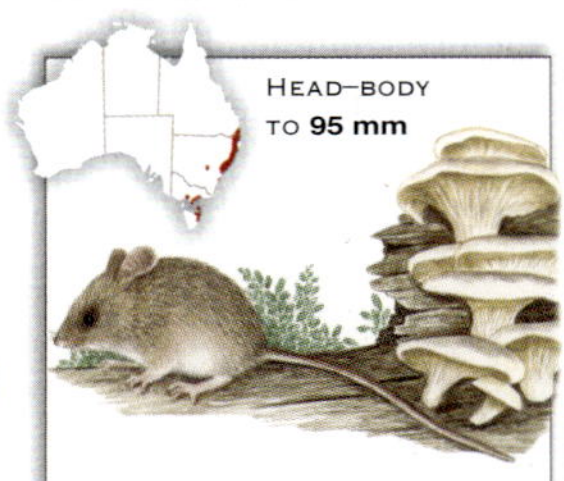

HEAD–BODY TO **95 mm**

New Holland Mouse **175**

RODENTS

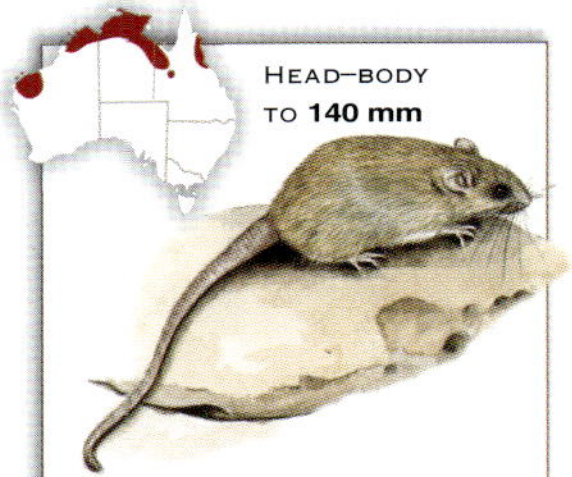
HEAD–BODY to 140 mm

Common Rock-rat **176**

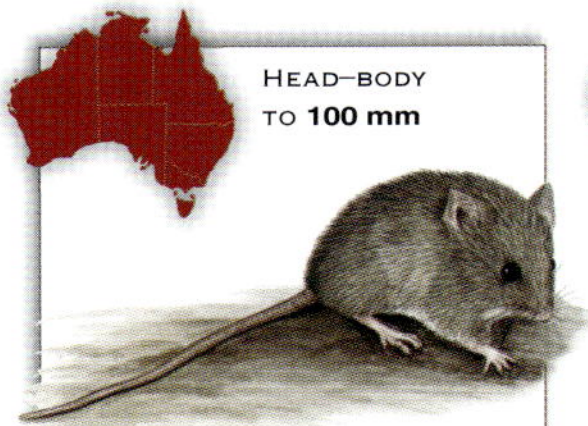
HEAD–BODY to 100 mm

House Mouse **181**

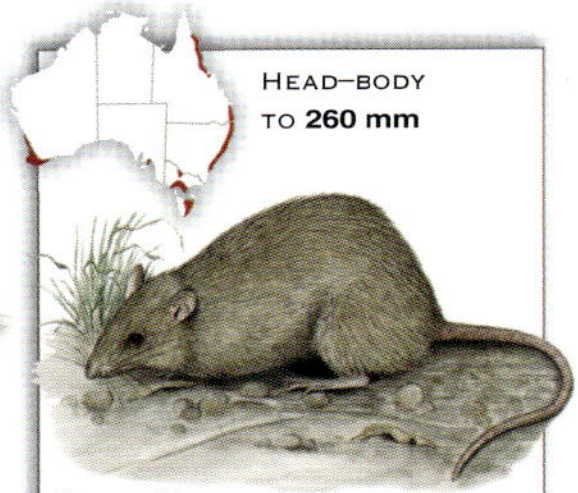
HEAD–BODY to 260 mm

Brown Rat **186**

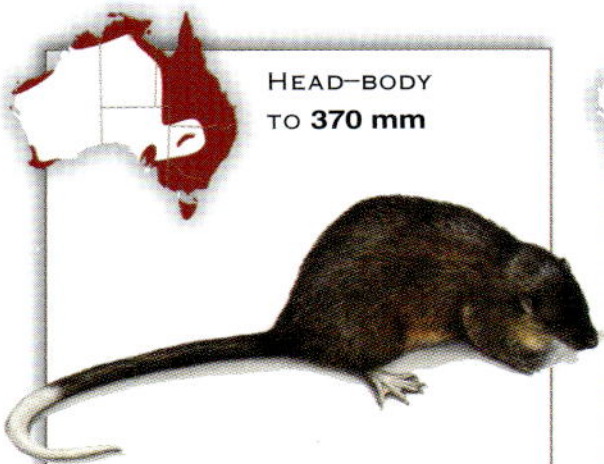
HEAD–BODY to 370 mm

Water-rat **177**

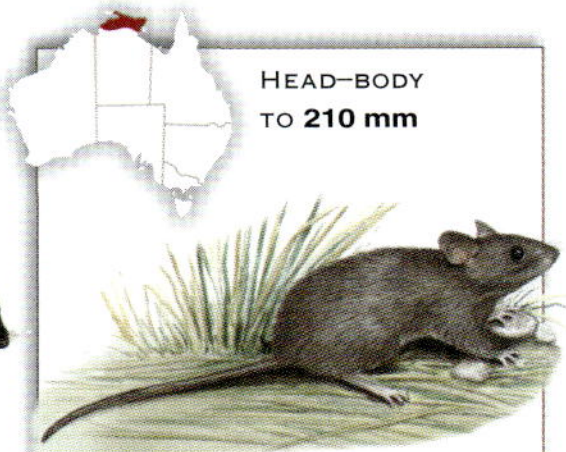
HEAD–BODY to 210 mm

Dusky Rat **182**

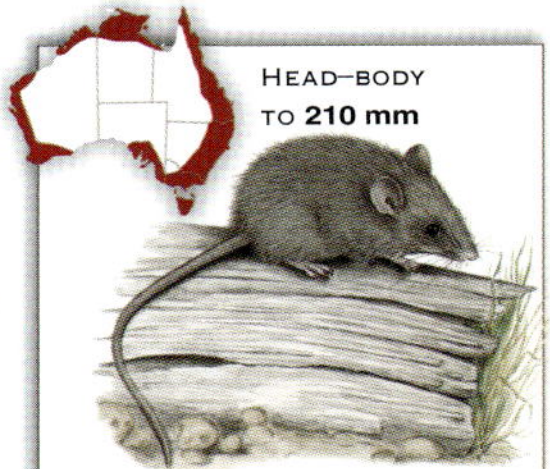
HEAD–BODY to 210 mm

Black Rat **187**

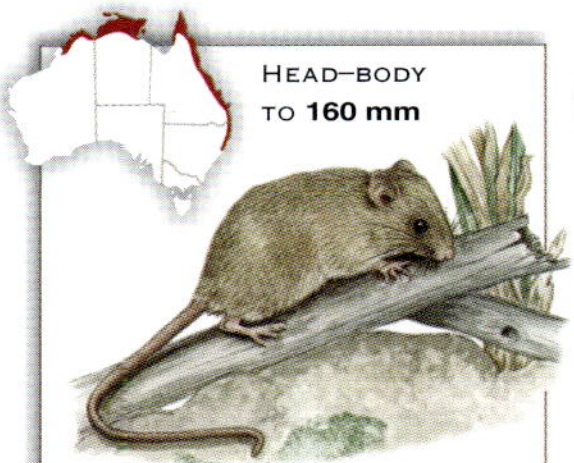
HEAD–BODY to 160 mm

Grassland Melomys **178**

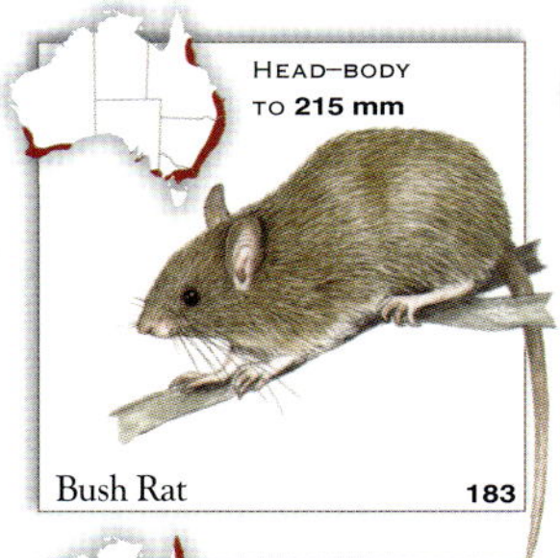
HEAD–BODY to 215 mm

Bush Rat **183**

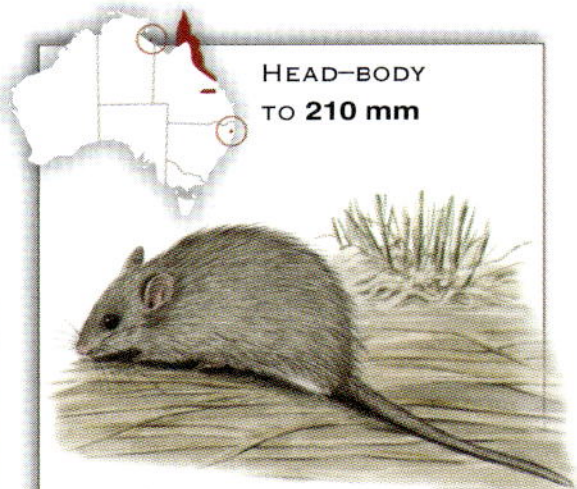
HEAD–BODY to 210 mm

Canefield Rat **188**

HEAD–BODY to 200 mm

Fawn-footed Melomys **179**

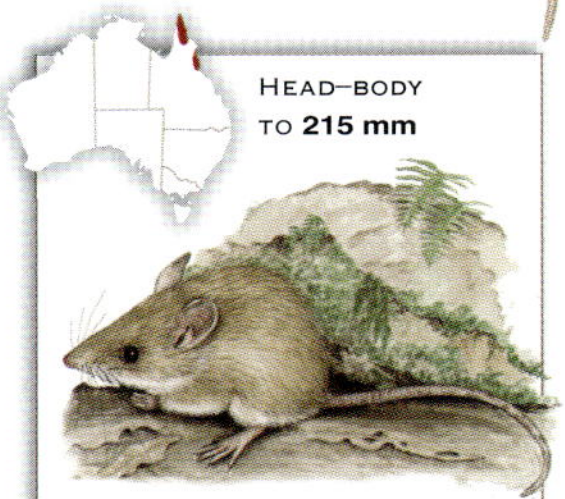
HEAD–BODY to 215 mm

Cape York Rat **184**

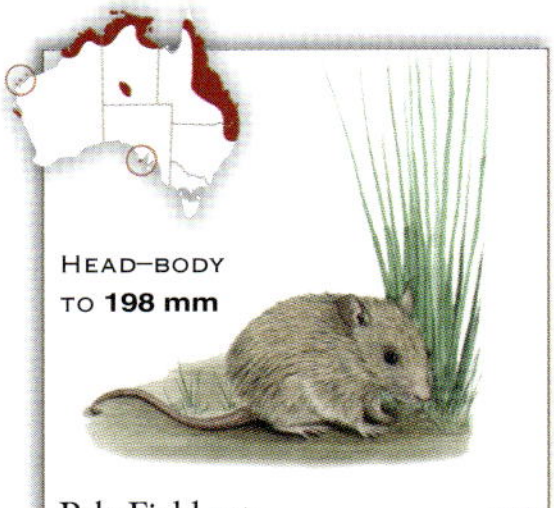
HEAD–BODY to 198 mm

Pale Field-rat **189**

HEAD–BODY to 385 mm

Giant White-tailed Rat **180**

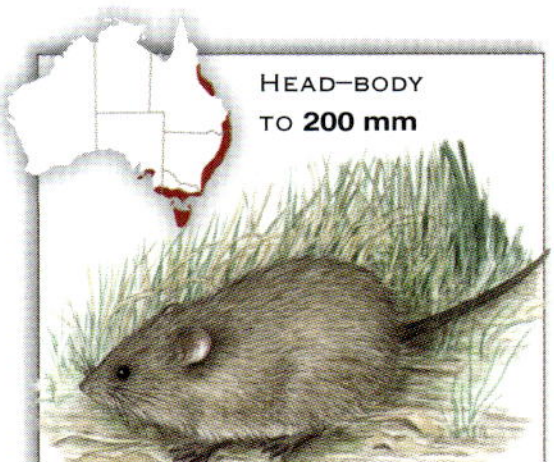
HEAD–BODY to 200 mm

Swamp Rat **185**

HEAD–BODY to 225 mm

Long-haired Rat **190**

DINGO

Dingo 191

MARINE MAMMALS

Australian Sea Lion 192

New Zealand Fur Seal 193

Australian Fur Seal 194

Leopard Seal 195

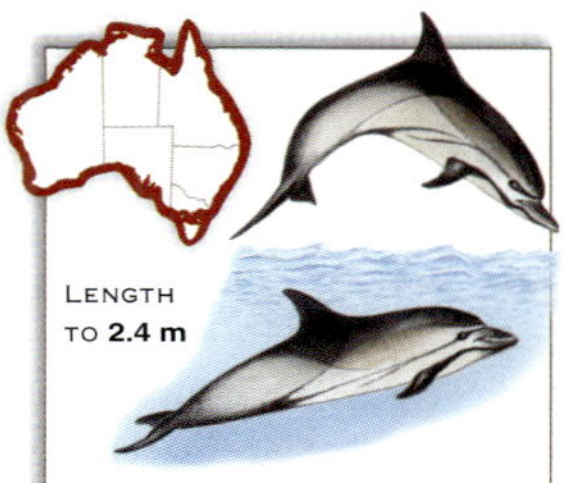

Common Dolphin 196

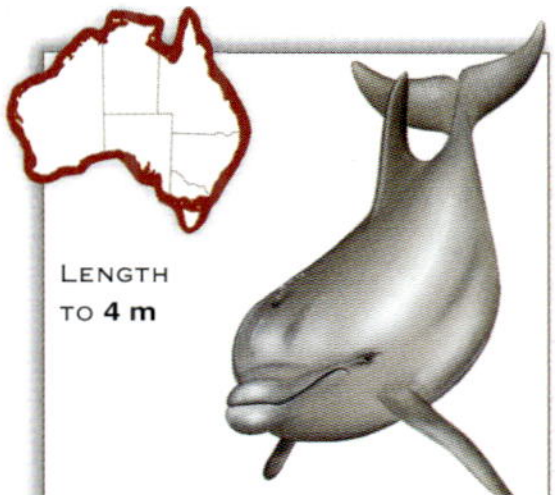

Bottlenose Dolphin 197

Dugong 198

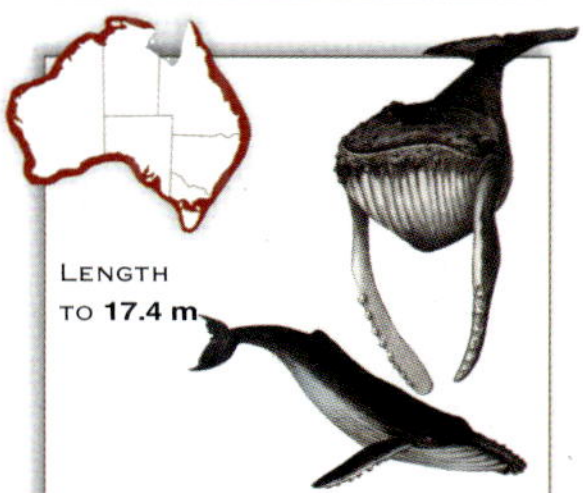

Humpback Whale 199

FAMILY **TACHYGLOSSIDAE** SPECIES *Tachyglossus aculeatus*

SHORT-BEAKED ECHIDNA

The short-beaked echidna is an egg-laying mammal with stout sharp spines covering its back and sides, a tubular snout with a long sticky tongue for gathering termites, small eyes and a rudimentary tail. Fur is dark brown to sandy and often obscures the spines of Tasmanian animals. The spines moult annually and are creamy-coloured with dark tips. Its short, powerful limbs have spade-like claws on the forefeet for digging and long claws on the hind feet for grooming. Males also have a small spur on the ankles of their hind legs. Males are larger than females. **Behaviour** Usually active between dawn and dusk, echidnas forage for up to 18 hours a day, and sleep under bushes, leaf litter, in hollow logs, rock crevices, rabbit or wombat burrows. In cold weather they become torpid to save energy, and in alpine areas hibernate for up to 28 weeks over winter, waking briefly every 2–3 weeks. Echidnas have a slow, rolling gait, and can swim well. They probe the ground with their snout and dig for ants with their forefeet, finding them by smell and by special receptors in the snout that detect the ants' tiny electrical signals. If threatened they freeze or curl up and eventually seek shelter or bury themselves on the spot. Generally solitary, they have overlapping home ranges of up to 70 ha. In the breeding season a group of up to 10 males will follow a receptive female in single file (sometimes for several weeks) until she stops and grasps a tree with her forelimbs. The males dig trenches next to her, but the strongest dislodges the other males and lifts her up to mate.

Pregnant females dig a nursery burrow 1–1.5 m long. **Development** Echidnas have a lifespan of up to 50 years and mate from June to September. Females develop a temporary abdominal pouch at the onset of the breeding season. About 2 weeks after mating they curl up and lay a single egg, about 15 mm long with a leathery skin, directly into the pouch. It hatches after 10 days and the newborn suckles from 2 milk patches next to the pouch, staying there until it begins to grow spines at 6–9 weeks. Still blind, it is left in the nursery burrow (with the entrance plugged) while the mother forages. She returns every 3–10 days and suckles it for another 4–5 months. **Diet** Termites, ants, worms, larvae and other invertebrates, crushed with the tongue. **Habitat** Most habitats from forests to deserts, to about 1400 m, except cleared farmlands, upland rainforests and fern gullies. **Traces** Conical diggings. Scats are long cylinders up to about 20 mm across and 50 mm long, light to dark brown, containing insect fragments. **Status** Low risk.

WEIGHT **2–7 kg** HEAD–BODY **300–530 mm** TAIL **70–90 mm**

FAMILY **ORNITHORHYNCHIDAE** SPECIES *Ornithorhynchus anatinus*

PLATYPUS

The platypus is an egg-laying aquatic mammal with a streamlined body, webbed feet and a flattened, paddle-like tail. It lacks external ears, but has a rubbery, duck-shaped bill lined with special receptors which are extremely sensitive to touch and the electrical signals emitted by aquatic invertebrates. Its fur is soft, dense and water-repellent, lustrous dark brown on the back, pale cream and sometimes reddish below. The feet have long, sharp claws. Males have a sharp, hollow, horny spur about 12 mm long on the inside of the ankles of their hind legs, connected to a venom gland in the groin, probably used against other males in the breeding season. Females lose the spur before maturity. Males are larger than females. **Behaviour** Active usually around dawn and dusk, and by day in cold weather, platypuses are generally solitary, living along rivers in home ranges of up to 3.5 km. Home ranges of adult males are exclusive, while those of females overlap. They have a waddling walk supported by the knuckles of the forelimbs and swim smoothly with a rolling dive, closing their eyes under water. Aquatic animals are caught or sifted from the river bed by the bill, stored in cheek pouches and ground between horny ridges after surfacing. Individuals use several camping burrows along the river bank, each about 2 m long, 80 mm high and 100–120 mm wide, with entrances submerged or hidden under a log. Breeding females dig a nursing burrow above water level, up to 20 m long, usually with a number of entrances and side branches, ending in a nest chamber lined with grass or leaves. They often groom on a favourite rock or log, and if disturbed will make a low growl. In cold climates they hibernate from May to September. **Development** Sexually mature at 2 years, they mate in the water from August to October. Females usually lay 2 eggs in the nest chamber, about 17 mm diameter with soft leathery skins, and curl around them until they hatch about 10 days later. Newborn feed for 3–4 months on milk suckled from ducts on the mother's abdomen. Young may be left in the nest with the entrance blocked for up to 36 hours while the mother forages. They are weaned at 17 weeks and live to 11 years or more. **Diet** Insects, molluscs, worms, shrimps, trout eggs and small vertebrates. **Habitat** Streams and lakes, from fast flowing highland streams to broad, slow-moving coastal waterways, in forests and rainforests to 1400 m. **Traces** Mud slides on stream banks, surface bubbles from submerged animals. Scats are greyish-brown pellets up to about 5 mm across and 10 mm long, comprising mostly fine silt particles, deposited around burrow entrances. **Status** Endangered in SA. Threatened by pollution, water diversion and habitat destruction.

WEIGHT **660–2600 g** HEAD–BODY **380–580 mm** TAIL **83–153 mm**

FAMILY **DASYURIDAE** SPECIES *Dasycercus cristicauda*

CREST-TAILED MULGARA

A small, robust, carnivorous marsupial of the central Australian deserts, the mulgara has fine soft fur, light sandy-brown with a dark grey base on the back, and greyish-white belly fur. The short tail has a fattened, reddish base and a striking crest of dark brown hairs along the terminal half. The conical head has large eyes and thin, sparsely-haired ears. The hind feet have 5 toes with furrowed pads. Males are larger than females. **Behaviour** Active mainly at night, mulgaras sleep in grass-lined nests in complex burrow systems, often with several entrances about 25 mm across, numerous vertical shafts and deep, branching side tunnels. Burrows are dug in the flats between sand dunes or on the slopes of high dunes. Individuals use several burrows within their home range and make well-defined paths from the burrow entrance to a nearby bush where piles of scats are deposited. Mulgaras live in overlapping home ranges varying from 1–14.4 ha. They are generally solitary, socialising briefly during the breeding season, and may be seen basking in the sun on cold days. They run with a bounding gait and sit upright on their hind legs if disturbed. Fast efficient killers, they skin their prey neatly, leaving the inverted skin behind. **Development** Sexually mature at 10–11 months, mulgaras probably continue growing throughout their lives, living to 7 years or more. They breed from May to July. Females develop a shallow pouch on either side of their 6 teats during the breeding season, and give birth to up to 6 young 35–42 days after mating. Newborn attach firmly to the teats and are dragged around for 55–60 days. They are then left in the nest until weaned at 3–4 months. **Diet** Large insects and small vertebrates. They can survive without drinking. **Habitat** Arid, sandy, inland deserts, gibber plains and spinifex grasslands. **Traces** Scats are up to 5 mm across and 20 mm long; strong-smelling; twisted at one end; containing bone, fur and insect fragments; and left as markers near burrow entrances and on rocks. **Status** Vulnerable. Threatened by grazing stock, competition and predation by feral animals.

WEIGHT **60–170 g** HEAD-BODY **125–220 mm** TAIL **70–130 mm**

FAMILY **DASYURIDAE** SPECIES *Dasyuroides byrnei*

KOWARI

A rat-sized carnivorous marsupial, the kowari is light grey-brown to sandy-brown above, usually with a darker stripe down the forehead, and greyish-white below. The terminal half of the tail has a brush of dark brown hairs. The pointed head has large eyes and thin, sparsely-haired ears. The limbs are quite long with 4 toes on the hind feet. Males are larger than females. **Behaviour** Active mainly at night, kowaris are ground-dwellers, running with a bounding gait and sitting up on their hind legs if disturbed. They sometimes bask in the sun on winter days, and become torpid for short periods when food is scarce. They dig burrows for daytime shelter, or modify those of other mammals such as the bilby or some rodents, and sleep in a nest lined with leaves and other vegetable matter. Some burrows have a number of entrances and individuals use several burrows, sometimes sharing with others. They occupy overlapping home ranges of several sq km, scent-marking the boundaries with urine, scats and secretions from a chest gland. Vocalisations include a loud, defensive, staccato chattering and a threatening hiss accompanied by vigorous tail movements. **Development** Kowaris have a lifespan of up to 6 years in captivity. Females first breed in their second year, mating from April to December, and copulating for several hours over a 3-day period. Up to 6 young are born 30–36 days after mating. A shallow pouch develops from skin folds before birth, leaving the mother's 6 teats exposed. Newborn are about 3 mm long and attach firmly to the teats, where they are partially enclosed in the pouch until about 30 days old. They then hang below the mother and are dragged around until detaching from the teats some 3 weeks later. The young are then left in the nest or cling to the mother's back while she hunts, until they are weaned at about 4 months. They may produce a second litter in the same year after the first litter becomes independent at 6 months. **Diet** Insects, birds, reptiles, amphibians, small mammals and carrion. They do not need to drink if the food is moist. **Habitat** Stony deserts with sparse vegetation. **Traces** Scats are about 5 mm across and up to 40 mm long, twisted and pointed at one end, strong-smelling, left as markers near burrows and on rocks. **Status** Vulnerable. Threatened by grazing, competition and predation by feral animals.

WEIGHT **70–140 g** HEAD–BODY **135–180 mm** TAIL **110–140 mm**

FAMILY **DASYURIDAE** SPECIES *Dasyurus geoffroii*

WESTERN QUOLL OR CHUDITCH

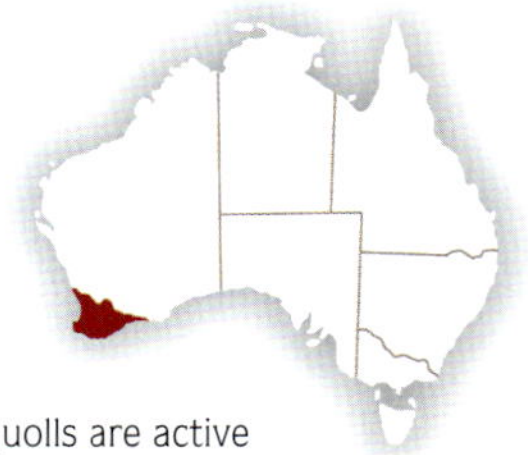

Since European settlement, the western quoll has disappeared from most of the Australian mainland and is now confined to southwestern WA. A cat-sized carnivorous marsupial, it has soft brown fur with conspicuous white spots above and is creamy-white below. Its tail is bushy with long black hairs towards the end and may have a few spots. The pointed head has large eyes and rounded ears trimmed with white fur. The hind feet have 5 toes with granular pads. **Behaviour** Western quolls are active at night and sometimes during the day at the height of the breeding season or in cold, wet weather. They sleep in grass or leaf-lined nests in hollow logs, rock piles, holes in termite mounds or in the burrows of other animals. They are usually solitary and use dozens of dens in a central core area of a home range averaging around 400 ha for males and 100 ha for females. Male home ranges overlap those of other males and several females. Most foraging takes place on the ground, although they occasionally climb small trees to catch prey or escape predators, clasping the trunk between their hind feet while gripping with the forefeet. Pregnant females construct a large burrow before giving birth, and share this with their offspring until they disperse in summer at 5–6 months old. **Development** Sexually mature at 12 months, western quolls have a lifespan of up to 5 years in captivity and around 3 years in the wild. Litters of 2–6 young are born from May to September with a peak in June–July, 17–18 days after mating, and suckle in the mother's moist, shallow pouch for 8–9 weeks. Barely able to crawl, and with little fur, the young have outgrown the pouch and are left in the den while their mother forages. By 16 weeks they are well-furred and are weaned by 22–24 weeks old. **Diet** Mammals, frogs, birds, lizards, insects, crustaceans and fruit. They rarely need to drink. **Habitat** Drier woodlands, mallee shrublands and sclerophyll forests. **Traces** Bone accumulations under rocks and in tree hollows. Burrows with a single entrance 100–200 mm across are often found at the base of a stump or under boulders. Scats are about 10 mm across and 30–40 mm long; light yellow to black; usually twisted at one end; containing fragments of insects, bones, fur or feathers; deposited conspicuously on rock piles or boulders. **Status** Vulnerable. Threatened by habitat loss and predation by feral animals.

WEIGHT **0.6–2.1 kg** HEAD–BODY **260–400 mm** TAIL **210–350 mm**

Family **Dasyuridae** Species *Dasyurus hallucatus*

Northern Quoll

The northern quoll is a carnivorous marsupial the size of a large rat. It is the smallest quoll. It has brownish-grey to brown fur with white spots on the back, but not on the tail, and cream to white belly fur. The head is long and pointed with large bulging eyes and large pointed ears. The hind feet have 5 toes with furrowed pads; the first toe lacks a claw. Males are larger than females. **Behaviour** Active mainly at night, northern quolls are good climbers although most of their time is spent on the ground. They are solitary and occupy home ranges averaging 35 ha in lowland savanna, although males expand their territory in the breeding season to up to 100 ha, overlapping the ranges of several other males and females. They sleep in dens in hollow tree trunks, hollow logs, crevices and holes in termite mounds. They are very aggressive and emit sharp screeches if threatened. During the breeding season each female is visited by up to 4 males per night. Males rove around regularly visiting females to monitor the onset of oestrous. Males are under great stress at this time, fighting other males in territorial defence and travelling large distances every night. **Development** Sexually mature at 10–11 months, they may live for 2–3 years although in the savanna most adult males die from stress-related illnesses soon after mating, and none survive to breed again. They breed in late June, giving birth to up to 8 young in July. Females have no true pouch although ridges of skin develop on either side of their 6–8 teats during the breeding season. The young remain firmly attached to the teats for the first 8–10 weeks, when they are still blind but well-furred. They are then left in the den while the mother forages, and are weaned at about 5 months. Most juvenile males and some young females leave the mother's home range at 6–8 months old. **Diet** Small mammals, reptiles, worms, insects, larvae, honey and soft fruits. **Habitat** Most treed habitats, favouring rocky areas in open forests, savanna and woodland; sometimes around camp grounds and houses. **Traces** Scats are about 10 mm across and 30 mm long; usually twisted and pointed at one end; containing fragments of insects, bones, fur and feathers; deposited on conspicuous high places on rock piles or boulders and at den entrances. **Status** Endangered. Threatened by habitat destruction, predation by feral animals and poisoning by cane toads.

Weight **300–950 g** Head–body **125–310 mm** Tail **125–310 mm**

FAMILY **DASYURIDAE** SPECIES *Dasyurus maculatus*

SPOTTED-TAILED QUOLL

A ferocious, cat-sized marsupial, the spotted-tailed quoll is the largest marsupial carnivore on the mainland. It has a powerful body with rich reddish-brown to olive-brown or chocolate fur, with white spots of various sizes on the back and tail, and pale creamy-yellow belly fur. A good climber, it has a long tail for balance, sharp curved claws, ridged pads on the feet, and an opposable clawless thumb on the hind foot for gripping branches. The face is relatively short with a squat, blunt muzzle. Males are larger than females. **Behaviour** Spotted-tailed quolls are active at night, although on cool winter days they sometimes hunt and bask in the sun. Mainly terrestrial, they are agile climbers and run with a bounding gait, otherwise their movements are slow and deliberate, sniffing regularly for signs of food. They climb trees to spot and pounce on prey, hunting and scavenging within a home range of 100–580 ha, often covering several kilometres in one night. Food is manipulated with the forepaws, and larger prey is killed by biting the back of the neck and head. They sleep in dens in hollow trees, hollow logs, caves, rock crevices or abandoned burrows. Dens are shared after mating by family groups, and defended aggressively by both sexes. Males fight for dominance during the mating season, uttering growls and staccato cries during courtship. When threatened they utter a piercing scream. Females make soft 'chh-chh-chh' calls to their young, and males sometimes bring food to the suckling mother. **Development** Sexually mature at 12 months, wild quolls breed at 2 years old. They have a lifespan of up to 5 years and breed from June to September in the north and April to August in the south. Copulation may last 8 hours. Females have a shallow, rear-opening pouch partially covering 6 teats. Up to 6 young are born some 21 days after mating and attach to the mother's teats for about 7 weeks. They are then left in the den while the mother forages, venturing out at about 14 weeks and becoming independent at about 18 weeks. **Diet** An opportunistic hunter and scavenger of birds, reptiles, insects and small to medium-sized mammals, particularly rabbits. **Habitat** Densely-vegetated areas in forests, rainforests, woodlands and coastal heaths. **Traces** Scats are often left in conspicuous places. They are 10–20 mm across and up to about 60 mm long; often twisted and pointed; dark to pale brown; containing fragments of bone, feathers, fur, reptile scales or insect remains. **Status** Endangered. Threatened by land clearing, logging, poisoned baits, predation and competition from feral animals, and poisoning by cane toads.

WEIGHT **2–7 kg** HEAD–BODY **350–760 mm** TAIL **340–550 mm**

FAMILY **DASYURIDAE** SPECIES *Dasyurus viverrinus*

EASTERN QUOLL

The size of a small cat, this carnivorous marsupial is of medium to slender build with dense soft fur. Two distinct colour variations exist, often in the same litter: either black with a brown belly or fawn with a white belly; both have white spots on the back but not on the tail. They have large rounded ears, relatively short legs and only 4 toes on the hind feet. **Behaviour** Active mainly at night, they hunt alone in open grasslands, among the undergrowth and among the lower branches of trees, climbing well and often standing erect on their hind legs sniffing the air. They sleep in grass-lined dens, often with several chambers, in short burrows, hollow logs, rock piles or dense vegetation. Usually solitary, they have overlapping home ranges, utilising and occasionally sharing a number of dens, particularly during the breeding season when social interactions increase and fighting often occurs between males. Males may travel over 1 km in a night, but females stay within a few hundred metres of their dens. Males change dens frequently but rarely share dens with other males. They utter a guttural growl when alarmed. **Development** Sexually mature at 12 months, eastern quolls breed from May to June. Females have no true pouch although ridges of skin develop on either side of the 6 teats during the breeding season. Up to 30 tiny young may be born 20–24 days after mating, but only those able to attach to a teat survive. A second litter may be produced if the first is lost early in lactation. Newborn are only about 6 mm long, attach firmly to the mother's teats, and are dragged around for about 10 weeks. They are then left in the den or are carried on her back if she changes dens, until weaned at 18–20 weeks, becoming independent at 20–29 weeks. **Diet** Mostly large insects and grubs; small mammals including mice, bandicoots and rabbits; reptiles; ground-nesting birds; fruits, seeds and grasses. They also scavenge the carcasses of large animals. **Habitat** Wet and dry sclerophyll forests, rainforests, heaths, scrubs and alpine areas, preferring dry grassland and forest mosaics bounded by agricultural land. **Traces** Scats are 10–20 mm across and up to about 60 mm long; twisted; containing bones, fur and feathers, or insect remains. **Status** Possibly extinct on the mainland; low risk in Tas.

WEIGHT **600–2000 g** HEAD–BODY **280–450 mm** TAIL **170–280 mm**

FAMILY **DASYURIDAE** SPECIES *Parantechinus apicalis*

DIBBLER

The size of a small rat, this carnivorous marsupial was considered extinct until it was rediscovered near Albany on the south coast of WA in 1967. The dibbler has brownish-grey, fairly coarse fur flecked with white above, giving it a freckled appearance. The belly fur is greyish-white tinged with yellow. It has a distinctive tapering, hairy tail and a white ring around the eye. **Behaviour** Dibblers are active mainly at night with an increase in activity around dawn and dusk. They are also thought to be active during the day, probably in cold, wet conditions. They forage among the dense leaf litter on the ground and often climb trees, moving quickly and easily along branches and among foliage, jumping between branches, licking nectar from flowers (particularly banksias) and catching insects. Prey is held and manipulated with the forepaws. Dibblers are thought to shelter in seabird burrows on the Boullanger and Whitlock Islands. Historical records describe nests of grasses and twigs hidden beneath vegetation on the ground and in the hollows of tree stumps. **Development** Sexual maturity is reached at 10–11 months. The breeding season extends from March–April with copulation lasting for several hours and a pair copulating more than once during the mating period. Females give birth to up to 8 young and carry them in a shallow pouch until they are too large; they are then left in the mother's nest while she forages. The young are dependent for 3–4 months, and juveniles first appear between September and November. Males sometimes die after breeding, although this does not occur every year.
Diet Insects, spiders and other invertebrates; reptiles; supplemented with plant matter including berries and nectar. **Habitat** Heaths and mallee-heath communities with sandy soils, dense ground cover and a substantial litter layer. **Status** Endangered. Threatened by habitat destruction and predation by feral animals.

WEIGHT **40–100 g** HEAD–BODY **140–145 mm** TAIL **95–115 mm**

Family **Dasyuridae** | Species *Pseudantechinus bilarni*

Sandstone Dibbler

A mouse-sized carnivorous marsupial, the sandstone dibbler has flecked greyish-brown fur above and pale grey belly fur with sandy patches behind its large, thin, crinkled ears. The tail is long and slender, sparsely-haired with visible scales. The head is long with large bulging eyes and a pointed muzzle with little fur on the sides. The feet have furrowed pads to help them grip when climbing. **Behaviour** Sandstone dibblers are mainly active at night, although in the winter they are often seen basking in the sun. They are agile climbers, but tend to spend most of their time on the ground foraging for insects. They sleep during the day in rock crevices, moving to humid vine thickets in the dry season in search of food. Males are more mobile than females. The majority of juveniles leave the maternal home range and disperse to a new area. **Development** The sandstone dibbler becomes sexually mature at 12 months. Mortality rates are high, and only about one quarter live to see a second breeding season. They breed from late June to early July, usually producing 4–5 young about 38 days later. Newborn attach firmly to the 6 teats on the mother's belly. Females have no true pouch and the young hang from their mother's teats as she moves around. After detaching from the teats they are left in the nest while she forages and are weaned at 4–5 months.
Diet Insects, larvae and small invertebrates.
Habitat Rugged sandstone escarpments with open eucalypt forests, vine thickets and closed forests. **Status** Low risk. Common in a limited habitat.

Weight **12–40 g** | Head–body **80–115 mm** | Tail **90–125 mm**

FAMILY **DASYURIDAE** SPECIES *Pseudantechinus macdonnellensis*

FAT-TAILED PSEUDANTECHINUS

A small carnivorous marsupial of the arid desert regions of Australia, the fat-tailed pseudantechinus derives its name from the swollen base of the tail where deposits of fat are laid down when food is plentiful. Poorly-nourished individuals use these fat reserves and consequently have a relatively thin tail. The fur is greyish-brown above and greyish-white below with light reddish-brown patches behind the ears. The head is long with a black-tipped, pointed muzzle, large eyes and ears. The hind feet are broad. **Behaviour** The fat-tailed pseudantechinus is active mainly at night, although individuals often emerge during the day in winter to bask in the sun, close to a rock crevice for escape. They sleep in deep rocky crevices or in burrows in termite mounds. In winter, and when food and water availability is low, they become torpid for several hours during the day to lower their body temperature and reduce their energy and water consumption. Unlike other dasyurids of a similar size, individuals maintain large long-term stable home ranges of around 1 ha for males and 0.3 ha for females. Male home ranges often overlap those of females. **Development** Sexual maturity is reached at the age of 10–11 months. Some males live long enough to breed twice, and some females survive to produce young in 3 breeding seasons. Births are timed to coincide with plentiful food supplies. They breed once a year in June–July in central Australia and in August–September in the western part of their range. Females have no true pouch although ridges of skin develop on either side of the 6 teats during the breeding season, and partially enclose the developing young. Up to 6 young are born 43–55 days after mating and attach themselves firmly to the teats where they remain, being dragged around by the mother until they detach and are able to be left in the nest while she forages. The young are weaned at about 14 weeks. **Diet** Insects, snails, other small invertebrates and occasionally small mammals (possibly carrion).
Habitat Largely confined to arid, rocky environments in central Australia.
Status Low risk.

WEIGHT **20–45 g** HEAD–BODY **95–105 mm** TAIL **75–85 mm**

FAMILY **DASYURIDAE** SPECIES *Sarcophilus harrisii*

TASMANIAN DEVIL

Largest of the carnivorous marsupials, the Tasmanian Devil is powerfully built with small eyes set in a short, broad head with strong jaws and well-developed teeth. The fur is black, usually with white patches on the neck and rump. The muzzle and small rounded ears are sparsely-haired, and they blush deep red if stressed. The limbs are short with strong claws. Fat is stored in the tail which becomes carrot-shaped. Males are larger than females. **Behaviour** Active mainly at night, they usually amble slowly along with an awkward lope, but can gallop quickly with both hind feet together. Young devils are more agile and are able to climb trees. They sleep in grass-lined dens in burrows up to 15 m long, with one or more entrances 200–300 mm diameter, sometimes shared by 2 or more animals. They also shelter in hollow logs, caves, rock piles and wombat burrows, and males make use of several dens. Devils are generally solitary with large overlapping home ranges of some 8–20 sq km. They are shy, opportunistic carnivores and inept killers, feeding mainly on carcasses, following well-defined trails to food sources, travelling around 8 km (occasionally up to 50 km) overnight, and gathering in groups to feed on large animals where they fight noisily over the remains. Devils establish dominance hierarchies and engage in ritualised aggression, standing on their hind legs and locking jaws, uttering loud growls, yells and screams. A sharp sneeze is used as a challenge and often becomes a fight. Devils produce a strong odour when under stress. **Development** Sexually mature at 2 years old, devils may live for 7–8 years. They breed mostly in March, and mate repeatedly over a 10 day period, with a dominant male guarding a female. Usually 2–3 young are born 21 days after mating. Each attaches to one of the 4 teats in the mother's shallow, backward-opening pouch, where they remain until fully-furred at 13–16 weeks. The young are then left in the den while the mother forages, and are weaned at 5–6 months, becoming independent at 9–10 months. **Diet** Mainly a scavenger feeding on carrion, including bones, supplemented by large insects, birds and other small animals. **Habitat** All major habitats, preferring sclerophyll forests, scrub and woodland. **Traces** Scats are grey, about 20 mm across and 30–60 mm long, pointed and twisted with fur at one end, full of bone and fur fragments, often deposited on walking tracks. **Status** Endangered. Threatened by the devil facial tumour disease.

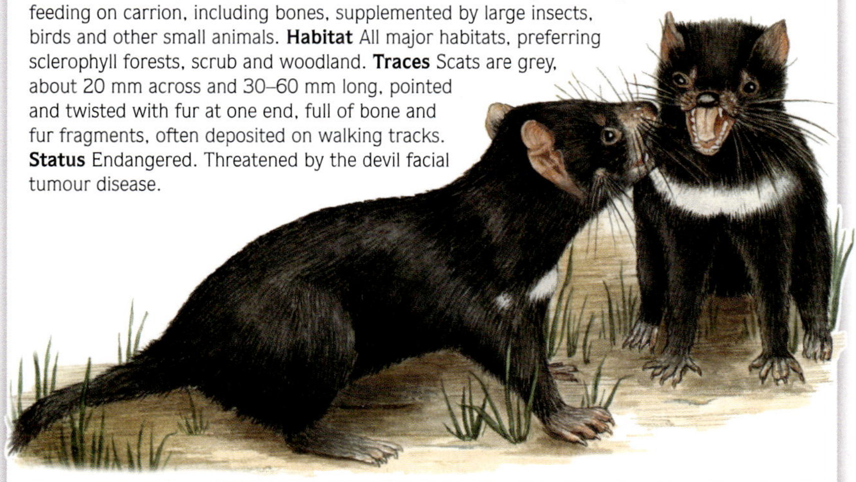

WEIGHT **7–10 kg** HEAD–BODY **500–710 mm** TAIL **240–310 mm**

FAMILY **DASYURIDAE** SPECIES *Antechinus flavipes*

YELLOW-FOOTED ANTECHINUS

One of the more colourful antechinuses, this mouse-sized carnivorous marsupial has a slate-grey head grading to orange-brown on the sides, belly, rump and feet; white patches on the throat and belly; a pale ring around the eye; and a black tail tip. Northern animals are more brightly coloured. The head is long and pointed with a black muzzle tip, bulging eyes and thin, crinkled ears. The forefeet have 5 long, well-clawed toes, the hind feet are very broad with 4 clawed toes and a small, opposing, clawed inner toe used to grip branches. Males are larger than females. **Behaviour** Active mainly at night, they are rarely seen during daylight hours, sleeping in a roughly constructed nest lined with leaves and other vegetation in hollow logs, rocky crevices, caves or buildings. They are highly arboreal, moving quickly and erratically with short bounds, running upside-down beneath branches, gripping with their broad feet, uttering high-pitched squeaks if threatened. They bulldoze through the leaf litter and pounce on their prey, killing quickly and efficiently with bites to the head and neck, leaving the skin of small mammals turned inside-out. The flattened head allows it to probe into small cavities looking for prey. In the breeding season males travel widely between nests. They are active during the day at this time and run around with a semi-erect penis, mating with all available females. Copulation may last for 12 hours. Some females survive to breed the following year, but all males die from stress-related illnesses soon after mating. **Development** Sexually mature at 10–11 months, they breed from July to September, depending on the climate. Each community has a synchronised 2 week breeding season. Each female stores the sperm from several dominant males during this period. The sperm then competes to fertilise her ova. Up to 12 young, from the strongest sperm of the strongest males, are born 23–26 days later. The number raised depends on the number of teats, which varies from 8–14. Females have no true pouch, although ridges of skin develop on either side of the teats 2 weeks before birth. The young attach firmly to the nipples and are dragged around for about 36 days. They are then left in the leafy nest or ride on the mother's back while she forages, and are weaned at 3–4 months. **Diet** Insects and other invertebrates, small vertebrates, flowers, fruit and nectar. **Habitat** Tropical rainforests to cool wet and dry sclerophyll forests and dry mulga woodlands, from sea level to 1000 m. **Traces** Scats are similar to the brown antechinus (page 35), although some may contain twists of fur or feathers. **Status** Low risk.

WEIGHT **21–80 g** HEAD–BODY **85–165 mm** TAIL **65–155 mm**

Family **Dasyuridae** Species *Antechinus stuartii*

Brown Antechinus

This small carnivorous marsupial has short, dense, coarse fur, dark greyish brown to orange-brown above, paler on the belly, and a sparsely-haired tail. It has a long head with bulging eyes and large, thin, crinkled ears with a notch in the margin. The forefeet have 5 strongly-clawed toes. The hind feet have 4 long, clawed toes and a small, opposing, clawless inner toe. Males are larger than females. **Behaviour** Active mainly at night, they usually stay on the ground, but may adopt an arboreal lifestyle in drier areas. They sleep in spherical nests of plant material in a hollow log, crevice or epiphytic fern. Nests are shared in winter months, with up to 18 animals huddling together. They sometimes bask in the sun, and become torpid for a few hours to conserve energy when food is scarce. Outside the breeding season they maintain exclusive foraging home ranges of about 1 ha and make use of several nests up to 500 m apart. As the breeding season approaches males leave their foraging ranges and group in all-male nests for most of the night. The scent gland on their chest enlarges, they become very agitated, highly aggressive and lose their caution, frequently emitting staccato chirps. Females continue to forage in their home ranges and visit some of the communal nests to find mates. Copulation lasts about 6 hours, and all males die from stress-related illnesses 5–10 days after the 2 week breeding season ends. Juvenile males disperse long distances from their mother's home range, while the females remain close by. **Development** Females may live for up to 3 years and both sexes are sexually mature at about 10 months. Mating occurs in August or September, and all the females in a particular population ovulate within a few days of one another. They copulate with several dominant males over a 2 week period. The sperm then competes to fertilise the ova. Between 6 and 10 young, produced from the strongest sperm of the strongest males, are born 25–31 days later. Females have no true pouch, although large ridges of skin develop on either side of their 6–10 teats before birth. The young remain firmly attached to the teats and are dragged around by the mother for 4–6 weeks. They are then left in the nest while the mother forages until weaned at about 3 months. **Diet** Cockroaches, beetles, spiders and other small invertebrates gleaned from the forest floor or fissures in tree bark; some small vertebrates, flowers and fruit. **Habitat** Rainforests and sclerophyll forests, preferring wetter habitats with dense undergrowth and deep leaf litter. **Traces** Scats are light to dark brown, about 2 mm across and up to 15 mm long, friable and packed with insect remains. **Status** Low risk.

Weight **17–71 g** Head–body **77–140 mm** Tail **65–110 mm**

FAMILY **DASYURIDAE** SPECIES *Antechinus swainsonii*

DUSKY ANTECHINUS

The dusky antechinus is a stocky, mouse-sized carnivorous marsupial with soft, dense fur, deep chocolate-brown to blackish-brown above and pale grey below. The head has a long pointed muzzle, small eyes and ears. The broad hind feet have 4 long, clawed toes and a small, clawless, opposing first toe for gripping branches. The forefeet have 5 toes with long, curved claws for digging. Males are larger than females. **Behaviour** Active both day and night, they sleep in shallow burrows often dug into creek banks, under logs or in cavities in logs or stumps. Tunnels are less than 1 m long and terminate in a spherical, leaf or grass-lined nesting chamber about 100 mm across. They are very active and fast-moving with short, jerky bounds, searching among leaf litter or digging for food. Although a ground-dweller, juveniles sometimes climb in the lower limbs of trees. In alpine areas they forage in tunnels beneath the snow. They maintain a home range, and females move their young between nests if disturbed. Males are solitary except during the mating season when they stop eating, become very aggressive towards other males and smear objects with secretions from their chest gland. Copulation is violent and may last 6 hours. Less than half of the females survive to the next breeding season, and all males die of stress-related illnesses within 3 weeks of mating. Juvenile males disperse abruptly from the maternal home range, while young females remain close by. **Development** Sexually mature at about 11 months, mating is highly synchronised in each population and takes place from May to August, or September in alpine areas. All females ovulate within 10 days of each other. Up to 10 young are born 28–35 days after mating, but only those able to attach to the mother's 6–8 teats survive. Females have no true pouch; ridges of skin enlarge around the teats before birth, forming a shallow pouch divided by a central ridge. The young remain firmly attached to the nipples and are dragged around until they are 7–8 weeks old. They are then left in the nest while the mother forages, venturing out at about 10 weeks. They are often carried on her back until independent at about 13 weeks. **Diet** Worms, beetles, centipedes, spiders, other invertebrates and some small vertebrates. **Habitat** Wet sclerophyll forests and woodlands with a dense understorey, rainforests, wet heaths and alpine heaths to 1800 m. **Traces** Scats are light to dark brown, 2–7 mm across and up to 30 mm long, twisted with fur or feathers, or packed with insect remains. **Status** Low risk.

WEIGHT **35–180 g** HEAD–BODY **90–190 mm** TAIL **70–130 mm**

Family **Dasyuridae** Species *Phascogale calura*

Red-tailed Phascogale

The red-tailed phascogale is a small, predominantly arboreal carnivorous marsupial distinguished by its long, colourful tail. The tail is rufous-brown with a black bushy tip. The body fur is ash-grey above and cream to white below. It has a pointed face, bulging eyes and large, thin, crinkled ears. The forefeet have 5 long, clawed toes, the hind feet have 4 long, clawed toes and a small, clawless, opposable inner toe. Males are larger than females. **Behaviour** They are predominantly nocturnal, emerging occasionally during the day. They sleep in cup-shaped nests of leaves and bark built in the skirts of grass trees, in the forks of trees, in hollow logs or in hollows in mature trees, particularly wandoo and rock oak. They hunt extensively on the ground, bounding along in a jerky manner, or walking carefully, stalking their prey. They are also skilful climbers and can leap 2 m between trees. Both sexes have anal glands that secrete a musky substance in the breeding season. The chest fur of males becomes stained with secretions from a chest gland during the breeding season, and this may be important in maintaining dominance hierarchies. Before giving birth, pregnant females force the males to disperse and all the males die within a few weeks from stress-related illnesses, probably induced by strenuous competition with other males during courtship and mating.

Development Sexual maturity is reached at about 9 months. Females live for up to 3 years, while males live to only 11.5 months. Mating takes place in July, and up to 13 young are born 28–30 days later. Only those able to attach to the mother's 8 teats survive. Females have no true pouch although ridges of skin develop on either side of the teats during the breeding season. Young are dragged around by the mother until they relinquish the teat at about 6 weeks. They are fully furred by 11 weeks, and are left in the nest until weaned at 3–4 months. **Diet** Insects and spiders, small birds, mammals, reptiles and possibly carrion. They can survive without drinking. **Habitat** Dry mature eucalypt forests and shrublands with an annual rainfall of 300–600 mm, preferably with a continuous canopy and numerous nesting hollows.

Traces Scats are 1–3 mm across and about 10 mm long, grey to yellow or brown, comprising insect fragments or narrow twists of fur or feathers.

Status Endangered. Threatened by land clearing, logging, bushfires, grazing stock and predation by feral animals.

Weight **38–68 g** Head–body **90–125 mm** Tail **120–145 mm**

FAMILY **DASYURIDAE** SPECIES *Phascogale tapoatafa*

BRUSH-TAILED PHASCOGALE

A semi-arboreal marsupial, the brush-tailed phascogale has steel-grey fur flecked with black above and pale cream below. The tail has a bushy tip with black hairs up to 40 mm long, and is not prehensile. The forefeet have 5 long, thin, clawed toes. The hind feet have 4 long toes with long, sharp claws, very flexible joints, and an opposing small, clawless first toe, enabling it to grasp branches and climb with ease. The head is long and pointed with large bulging eyes and long ears. Males are larger than females. **Behaviour** Active at night, they are fast, agile climbers, running beneath branches, leaping between trees and rapping their feet on the trunk if threatened. They bite off the bark and extract insects with their long fingers. By day they sleep in nests in the ground or in globular bird nests, using up to 40 different sites annually and often changing daily. Nursery nests are lined with leaves, shredded bark, feathers and fur, and built in large tree hollows with an entrance 25–40 mm across. They are generally solitary, although pairs sometimes share nests in the breeding season. Females have exclusive home ranges of 20–70 ha, sometimes shared with their female offspring. Juvenile males disperse when they leave the nest and establish overlapping home ranges of more than 100 ha. In the breeding season males travel long distances searching for females, and their chests become stained with yellow secretions. All males die a few weeks after mating due to stress-related illnesses. **Development** Sexual maturity is reached at about 8 months, and females may live to 3 years. Mating occurs in May and July in tree hollows, and copulation may last several hours. Females develop a shallow abdominal pouch in the breeding season from enlarged ridges of skin around the teats. More than 8 young may be produced some 30 days after mating, although only those able to attach to the mother's 8 teats survive. The young are carried around until they detach from the teats at about 7 weeks. They are then left in the nest while the mother forages and are gradually weaned between 14 and 25 weeks. **Diet** Insects, spiders, centipedes, small vertebrates and nectar. **Habitat** Wet and dry open sclerophyll forests and woodlands, occasionally in rainforests, heaths and swamps, preferring ridges and rocky slopes to 1500 m. **Traces** Scats are deposited around dens, on tree branches and logs. They are about 5 mm across and up to 40 mm long, pale yellow to blackish-brown, full of insect fragments or twisted with feathers and fur. **Status** Low risk nationally, but endangered in SA and vulnerable in the NT and NSW. Threatened by land clearing, logging, competition with honey bees for tree hollows, and predation by feral animals.

WEIGHT **105–310 g** HEAD–BODY **148–260 mm** TAIL **160–235 mm**

Family **Dasyuridae** Species *Planigale gilesi*

Paucident Planigale

A tiny, desert-dwelling carnivorous marsupial, the paucident planigale has soft, dense fur, cinnamon-grey with a black base above and olive-buff below. The head is flattened and triangular with small rounded ears and protruding eyes. The tail tapers to the tip, becoming carrot-shaped in good seasons when the base is swollen with stored fat. The limbs are adapted for climbing in soil cracks with short, broad, 5–toed feet with granular pads. It has only 2 premolar teeth (other planigales have 3 premolars). Males are larger than females. **Behaviour** Active mainly at night, paucident planigales are ferocious hunters and efficient killers, climbing among grass and shrubs and slithering through the ground litter searching for insects. They walk with their hind feet swinging outside the body line, keep low to the ground, and run with a scurrying gait taking short leaps and bounds. During daylight hours they sleep in nests in hollow logs, under bark, in grass clumps or short burrows, some with a number of side tunnels and sleeping chambers. In winter they share nests for warmth, bask in the sun, and become torpid for short periods when food is scarce. Individuals have shifting home ranges and males may establish dominance hierarchies. They are very mobile for most of the year, travelling more than 1 km during the night; but during the breeding season females stay within a small home range and are more sedentary. Vocalisations include an aggressive 'chh-chh' or 'ca-ca', a high-pitched twittering when stressed, and a clicking sound during courtship. **Development** They have an extended breeding season from July to February. Copulation may last more than 2 hours, and females can produce 2 litters per season of 6–8 young, usually born 16 days after mating. Newborn attach firmly to one of the 12 teats in the mother's rear-opening pouch, where they remain for about 37 days. Thereafter they are suckled in the nest until 65–70 days old. They have a lifespan up to 5 years, but few live more than 2 years. **Diet** Invertebrates including beetles, spiders, cockroaches and slaters, and occasionally reptiles and small mammals. **Habitat** Arid sandplains, floodplains and creek beds with cracking clay soil and a cover of woodlands, grasslands or sedges. **Traces** Scats are small, twisted, less than 3 mm across, friable and full of insect remains. **Status** Low risk.

Weight **5–16 g** Head–body **60–85 mm** Tail **55–72 mm**

FAMILY **DASYURIDAE** SPECIES *Planigale maculata*

COMMON PLANIGALE

A small, ferocious carnivorous marsupial, the common planigale is cinnamon to grey-brown above, sometimes flecked with white; pale brown to white below with a white chin. The head is triangular and flattened with small eyes and large ears with notched margins. The thin tail is shorter than the head and body. **Behaviour** Active mainly at night, common planigales are fierce hunters and agile climbers, preying on small animals almost as big as themselves. During the day they sleep alone or with others in a saucer-shaped nest lined with leaves, grass or shredded bark. Nests are built in crevices, hollow logs, beneath bark or under rocks. They bask in the sun on cold days and become torpid for short periods when food is scarce. In the breeding season males fight frequently and scent-mark areas with their chest gland, probably establishing dominance hierarchies which give mating preferences to the largest. They are adept climbers and freeze if disturbed. **Development** Sexually mature at about 10 months, both sexes may breed for at least 2 years. The breeding season extends from October to January in eastern Australia with peaks in spring and summer, and lasts all year in the Northern Territory with several litters per year. Females have 8–13 teats in a rear-opening pouch, and give birth to 4–15 young 19–20 days after mating. Newborn attach firmly to the teats for about 21 days, vacating the pouch at 45 days, and remaining in the nest until weaned at about 70 days. **Diet** Insects and small vertebrates, some as large as themselves. **Habitat** Wet sites in rainforests, sclerophyll forests, heathlands, shrublands, marshlands, grasslands and rocky areas with surface cover, usually near water. **Traces** Scats are small twisted cylinders, about 1 mm across and 1–8 mm long, friable, packed with insect fragments. **Status** Low risk nationally but vulnerable in NSW. Threatened by land clearing and predation by feral animals.

WEIGHT **6–22 g** HEAD–BODY **65–100 mm** TAIL **51–95 mm**

Family **Dasyuridae** | Species *Planigale tenuirostris*

Narrow-nosed Planigale

This tiny carnivorous marsupial derives its name from its slender flattened snout, used to probe into cracks or litter in search of prey. The silky fur has a black base with brown tips giving the back a russet-brown appearance flecked with black, and a faint dark stripe down the forehead. The belly fur is olive-buff to white. The ears are large with a number of folds. It has a short, thin tail and the hind legs have great flexibility allowing them to climb easily and investigate narrow crevices. **Behaviour** Active mainly at night, narrow-nosed planigales spend much of their time scurrying around in deep cracks in the ground, occasionally climbing the vertical sides, standing upright sniffing the air. They are ferocious hunters, killing insects as large as themselves and holding food with their forepaws. In winter they huddle together in nests of dried grass and bask in the sun, becoming torpid for short periods if food is scarce. They have shifting home ranges and may cover more than 500 m overnight foraging for food. Males travel widely during the breeding season, probably looking for females to mate with, and scent-mark areas by rubbing with their chest gland. They fight frequently, make short 'tsst, tsst' calls to attract females, and may establish dominance hierarchies giving the largest individuals mating preferences. Vocalisations are often loud and include an aggressive 'chh-chh-chh', a defensive 'ca-ca-ca' and high-pitched twittering when stressed. **Development** The breeding season extends from early August to January and females can produce 2 litters of 6–8 young per season. Young are born 19 days after mating, are 3 mm long and attach firmly to the 10–12 teats in the mother's rear-opening pouch. They detach from the teats at about 40 days and are left in the nest while the mother forages for food. Their eyes open at about 51 days and they are suckled for about 95 days. They grow throughout their lives and may live for 2–3 years. **Diet** Insects, spiders, and other arthropods, possibly small lizards. They do not need to drink. **Habitat** Sandplains, floodplains and creek beds with cracking clay soils and a cover of woodlands, grasslands or sedges. **Traces** Scats are similar to the common planigale (page 40). **Status** Low risk.

Weight **4–10 g** | Head–body **44–75 mm** | Tail **48–72 mm**

FAMILY **DASYURIDAE** SPECIES *Antechinomys laniger*

KULTARR

The kultarr is a mouse-sized carnivorous marsupial of the arid and semi-arid regions of Australia The fur is fawn-grey to sandy-brown above and white below. There is a dark ring around the eye and the forehead is slightly darker. The tail is very long and thin with a bushy tip of long darker hairs. The head is large with bulging eyes and very large ears. The hind legs are very long with only 4 toes. Males are large than females. **Behaviour** Active only at night, kultarrs are terrestrial animals, moving around slowly on all fours, but bounding from the hind legs onto the forefeet when moving quickly. They have great manoeuvrability, using their forefeet as pivots to change direction and escape predators. They frequently stand up on their hind legs to look around, especially if disturbed. During the day they shelter in hollow logs and tree stumps, under logs and rocks, beneath spinifex and saltbush tussocks, in cracks in the soil, and sometimes make use of the abandoned burrows of other small animals. They sometimes dig shallow burrows themselves to escape the summer heat, concealing the entrance with grass. They are generally solitary, only coming together during the mating season. When food is scarce they become torpid, slowing down their metabolism to save energy and water. **Development** Sexually mature at 8 months, kultarrs breed from July to January, although the timing depends on geographical location and the availability of food. Females develop a crescent-shaped pouch during the breeding season and may produce several litters in a good season, each with up to 6 young. Females give birth to tiny undeveloped young 12 days after mating. They crawl into the mother's shallow, rear-opening pouch and firmly attach to her 6 or 8 teats, staying there until they are about 30 days old and 25 mm long. They are then left in the nest or ride on the mother's back while she forages for food until weaned and independent at about 3 months old. **Diet** Insects, spiders and other small invertebrates. **Habitat** Desert plains, arid and semi-arid grasslands and scrublands, mallee woodlands. **Status** Low risk nationally, but endangered in NSW and rare in Qld. Threatened by habitat loss and predation by feral animals.

WEIGHT **20–40 g** HEAD–BODY **70–100 mm** TAIL **100–150 mm**

Family **Dasyuridae** Species *Sminthopsis crassicaudata*

Fat-tailed Dunnart

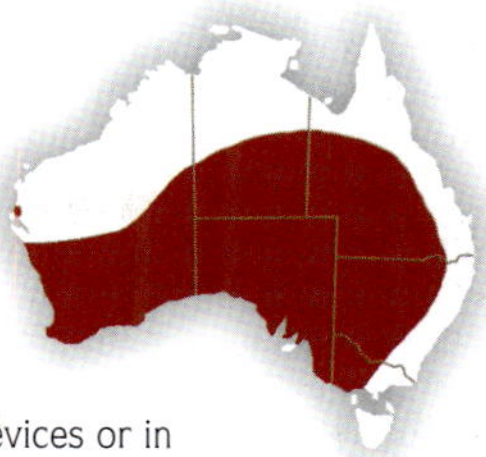

A widespread, mouse-sized carnivorous marsupial, the fat-tailed dunnart is brownish-grey above and light grey to white below. The fur has a dark base and sandy-grey tips. It has a pointed head with large bulging eyes and very large ears. Fat is stored in the short tail, which resembles a small carrot in well-nourished individuals. Males are larger than females. **Behaviour** Active only at night, they sleep by day in cup-shaped nests of dried grass, 70–100 mm across, beneath logs, rocky crevices or in deep cracks in the soil in arid areas. Solid spherical nests are sometimes built. In cold weather they often huddle together in communal nests to keep warm, and may be seen basking in the early morning sun. They become torpid for a number of hours if food is short, dropping their body temperature to conserve energy and utilising fat reserves in the tail. They are generally solitary and hunt for invertebrates on the ground, usually in open areas. They have large, drifting home ranges that change according to the food supply. **Development** Sexual maturity is reached at 5 months, although females do not breed in the year of their birth. Most females live for less than 18 months, while few males live more than 15 months. The breeding season extends from July to February with birth peaks in August and September, coinciding with the seasonal abundance of food. Two litters are usually raised in a year. The gestation period is very short, and 8–10 young are born only 13–16 days after mating. They attach firmly to the 8–10 teats in the mother's well-developed circular pouch with a central opening, detaching at around 43 days and leaving the pouch permanently at about 60 days. The young are then left in the nest while the mother forages, and are carried on her back if she moves between nesting sites until they are weaned at about 10 weeks. **Diet** Insects, earthworms and other small invertebrates; rarely small vertebrates such as frogs. They can exist without drinking. **Habitat** Moist coastal to arid inland areas with open woodlands, low open shrubland, sparse tussock grasslands, gibber plains and farmland; from sea level to 400 m. **Traces** Scats are small, twisted, 2–4 mm across and up to about 15 mm long, friable, packed with insect fragments, deposited in piles near nest sites. **Status** Low risk.

Weight **10–20 g** Head–body **60–90 mm** Tail **40–70 mm**

FAMILY **DASYURIDAE** SPECIES *Sminthopsis granulipes*

WHITE-TAILED DUNNART

A small, delicate, carnivorous marsupial, the white-tailed dunnart has long, soft fur with a blue-grey base and dark brown tips on the back, giving it an overall light fawn to grey colour. The tail is white with a thin dark brown stripe along the upper surface. The head is pointed with a narrow muzzle, large bulging eyes and large ears with notched margins. It has white paws and fine, even, rasp-like granulations on the pads on the soles of the forefeet and hind feet. **Behaviour** Although common in its habitat, the white-tailed dunnart has been little studied. It is known to be active at night, when it hunts for insects. Dunnarts have a high metabolic rate and when food is short they fall into a state of torpor, lowering their temperature and metabolic rate to conserve energy. **Development** The breeding season of the white-tailed dunnart is thought to extend from June to October. Females have a shallow circular pouch with a central opening and the tiny, undeveloped newborn crawl into the pouch and attach to the mother's teats. The young are probably weaned in September–October. **Diet** The white-tailed dunnart is an opportunistic feeder, eating available invertebrates including moths, ants, wasps, beetles, weevils, termites, cockroaches, horseflies, other insects and their larvae, spiders and centipedes. **Habitat** Low shrubland vegetation, sparse mallee with shrubby ground cover, mostly on sand.
Status Low risk.

WEIGHT **18–37 g** HEAD–BODY **69–88 mm** TAIL **56–66 mm**

Family **Dasyuridae** | Species *Sminthopsis hirtipes*

Hairy-footed Dunnart

A small, mouse-sized carnivorous marsupial of the drier regions, the hairy-footed dunnart is yellowish-brown to grey-brown above and light grey to white below. The base of the tail is slightly swollen and becomes fatter in good seasons. The feet are broad and long (16–19 mm) with fat toes. They have granular soles fringed with long, fine silvery hairs extending beyond the feet. The hairs are thought to help them walk on soft sand. The head is long with a narrow pointed muzzle, large ears and large bulging eyes. **Behaviour** The hairy-footed dunnart has been little-studied, partly because of the unsuccessful collecting techniques used in the past. It is active at night, and individuals have been found in deep burrows probably made by hopping-mice, and in abandoned bull-ant nests. Dunnarts have a high metabolic rate and when food is scarce they become torpid, lower their metabolic rate to conserve energy and use fat reserves stored in the tail. **Development** Lactating females have been found in February and April, suggesting spring and summer breeding seasons, although it is likely that the breeding season relates to rainfall, with population increases after sufficient rain. Females have 6 teats enclosed in a circular pouch with a central opening. Hairy-footed dunnarts have a lifespan of at least 3 years in the wild. **Diet** Insects and their larvae, and small lizards if they are available. **Habitat** Sandy arid to semi-arid low open woodlands, shrublands, heaths and hummock grasslands on plains or sand dunes. **Status** Low risk nationally, but rare in SA. Threatened by land clearing and predation by feral animals

Weight **13–20 g** | Head–body **70–85 mm** | Tail **70–100 mm**

FAMILY **DASYURIDAE** SPECIES *Sminthopsis leucopus*

WHITE-FOOTED DUNNART

A slightly-built, mouse-sized carnivorous marsupial, the white-footed dunnart has soft, light grey fur on its back, becoming darker on the rump and pale grey to white belly fur. It has a long pointed muzzle, large bulging eyes, large rounded ears, a scaly tail with sparse, coarse hairs and very narrow hind feet with furrowed pads to help it grip when climbing. Males are larger than females. **Behaviour** Active mainly at night, white-footed dunnarts are very fast on the ground and good climbers. They sleep in nests lined with roughly-shredded bark in tree hollows, rotting logs, wood piles, grass clumps, rock crevices, under bark sheets and in burrows. Generally solitary, they live in small home ranges about 100 m long for males and about 80 m long for females. Females have exclusive, non-overlapping home ranges, while those of males overlap. Some males travel more than 1 km in a night looking for new sources of food. Males do not survive to breed a second year and seem to disperse within 1 month of the young becoming independent. Prey is held in the forepaws to be eaten. **Development** Mating occurs in July and August, with births in late August and September. Most individuals have only one short breeding season in their lives. Up to 10 young, each about 3 mm long, are born and attach firmly to the mother's 8–10 teats, completely enclosed in a circular pouch with a central opening. They remain there for about 8 weeks. The young are then suckled in the nest for about 1 month. **Diet** A range of ground-dwelling invertebrates including insects and their larvae, small reptiles and possibly small mammals. **Habitat** Woodlands and open forests, wet heath, sedgeland, tussock grassland and coastal scrub; from sea level to 400 m; with an annual rainfall of around 600–1000 mm. **Traces** Scats are small, twisted, about 2–3 mm across and up to 15 mm long, very friable and full of insect fragments, deposited in small piles in the nest. **Status** Vulnerable in NSW. Threatened by land clearing, grazing, inappropriate fire regimes and predation by feral animals.

WEIGHT **20–35 g** HEAD–BODY **65–120 mm** TAIL **45–105 mm**

FAMILY **DASYURIDAE** SPECIES *Sminthopsis macroura*

STRIPE-FACED DUNNART

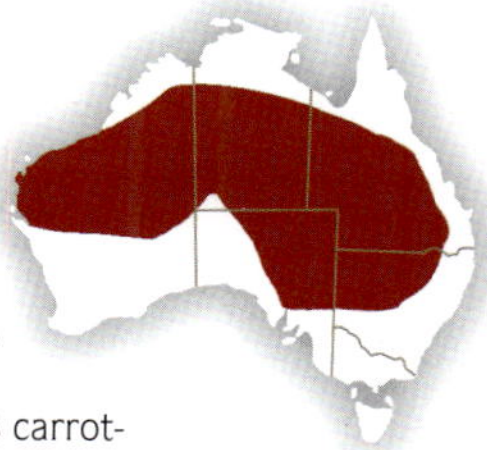

This widespread, small carnivorous marsupial is found over vast areas of the arid and semi-arid regions of Australia. The fur is brownish-grey to dark brown flecked with light grey above and pale grey to white below. It has a prominent dark stripe down the forehead. The head is long and pointed with large bulging eyes and large ears. The tail is long, sparsely-haired with a fattened base in well-nourished individuals, and tapers to a point. Fat is stored in the tail when food is plentiful and becomes carrot-shaped. **Behaviour** Little studied, they are known to be active at night, sheltering from the hot sun during the day in nests made from grasses and leaves. The nests are found under rocks in hollow logs, in dense vegetation, tussock grasses, in cracks in the ground or in burrows. In cold conditions or when food is short they enter a state of torpor for a few hours, lowering their body temperature to that of the surroundings, conserving energy and using fat stored in the tail. They hunt on the ground, walk slowly with a quadrupedal gait and run bipedally. **Development** Sexual maturity is reached at 4–5 months in females and at about 9 months in males. Mating takes place from June to February, with most births occurring in July, August and October to January. The young are born 11–12 days after mating and firmly attach to the 8 teats in the mother's well-developed, centrally-opening, circular pouch. They detach from the teats and leave the pouch at about 40 days and are then left in the nest while the mother forages until they are weaned at about 70 days. The short gestation period means that females can raise 2 litters of 6–8 young in good seasons. **Diet** Insects and other small invertebrates; small lizards, mice and other small vertebrates. They can probably survive without drinking. **Habitat** Arid and semi-arid low shrubland, woodland, spinifex and tussock grassland. **Status** Low risk nationally but vulnerable in NSW. Threatened by land clearing, grazing and predation by feral animals.

WEIGHT **15–25 g** HEAD–BODY **70–100 mm** TAIL **80–110 mm**

FAMILY **DASYURIDAE** SPECIES *Sminthopsis murina*

COMMON DUNNART

Although it is capable of rapid population growth in good years, this small insectivorous marsupial has a patchy distribution. It is slate grey above and white below. It has a long, pointed head with bulging eyes, large ears and a slender muzzle. The hind feet are less than 3 mm wide with a very small first toe. The tail is slender, sparsely-haired with visible scales. Males are larger than females. **Behaviour** Common dunnarts are active mainly at night, and sleep in cup-shaped nests, 70–100 mm wide, of dried grass and leaves constructed in hollow logs, grassy tussocks, natural crevices or grass-trees. In cold weather when food is scarce they become torpid, lowering their body temperature to conserve energy. Males become very aggressive during the breeding season, often fighting over females who attract them by a repetitive 'chit-chit-chit' call. Females may survive to breed a second time, but most males probably die from stress-related illnesses soon after the breeding season has ended. **Development** Sexual maturity is reached at about 6 months, and the breeding season extends from early September to late March. Females mate soon after weaning their first litter, and most females probably produce 2 litters of 8–10 young each breeding season. Young are born only 12.5 days after mating, and attach firmly to the mother's 8–10 teats, completely enclosed in a well-developed circular pouch with a central opening. They develop rapidly, vacating the pouch at about 34 days, and are then left in the nest while the mother forages. They are weaned at 60–65 days and reach adult size by about 5 months. The pouch regresses after weaning. **Diet** Spiders, beetles, caterpillars, cockroaches and other invertebrates. **Habitat** Open forests and woodlands, mallee and heath; to an altitude of 400 m; with an annual rainfall of 250–850 mm. Maximum population densities are reached in areas regenerating after fire. **Traces** Scats are deposited in large numbers near nest sites. They are small, twisted, about 2–4 mm across and up to 15 mm long, friable, containing fine insect fragments. **Status** Low risk. Widespread with patchy distribution and isolated populations.

WEIGHT **10–28 g** HEAD–BODY **60–105 mm** TAIL **68–100 mm**

Family **Dasyuridae** Species *Sminthopsis virginiae*

Red-cheeked Dunnart

A small carnivorous marsupial of northern Australia and New Guinea, the red-cheeked dunnart has short, spiky fur; blackish-brown flecked with white above; white, yellow or rufous below. It has reddish cheeks and a distinct blackish stripe down the forehead. The tail is narrow, sparsely haired and scaly. The hind feet are narrow with 4 long toes and a small first digit. The head is long and pointed with large bulging eyes and large rounded ears. **Behaviour** Red-cheeked dunnarts are active mainly at night and sleep during the day in a nest on the ground, under logs or under cover of dense vegetation. Shortly before giving birth pregnant females build a shallow, saucer-shaped nest of shredded leaves and grass hidden under dense grass or pandanus leaves. They are solitary except during the breeding season when females attract males by calling. This elicits similar searching and calling behaviour in males. **Development** Red-cheeked dunnarts become sexually mature at about 7 months and continue to grow throughout their lives. They are capable of breeding throughout the year, although this has not been verified in the wild. Females with suckling young have been found in March and July–November. Females may produce 2 litters per season of up to 8 young, born about 15 days after mating. Newborn attach firmly to the mother's 8 teats and are fully-enclosed in a well-developed circular pouch with a central opening. The young are carried in the pouch for about 55 days then left in the nest while the mother forages for food. They are weaned and independent at about 90 days. **Diet** A range of small invertebrates including insects and spiders, small lizards and other small vertebrates. **Habitat** Savanna woodlands. **Traces** Scats are thin, twisted, up to 5 mm across and 20 mm long, friable and full of fine insect fragments, deposited around nest sites. **Status** Low risk; sparsely distributed.

Weight **18–60 g** Head–body **90–135 mm** Tail **90–135 mm**

FAMILY **MYRMECOBIIDAE** SPECIES *Myrmecobius fasciatus*

NUMBAT

The numbat is the only marsupial adapted to feeding on termites. Like the echidna it has poorly-developed teeth and a long, sticky, cylindrical tongue that flicks rapidly in and out of the long snout to collect termites. The fur is reddish-brown above becoming darker on the rump and has a number of white transverse stripes. The belly is pale grey to white. A dark stripe runs across the eye from ear to mouth. It has a white patch below the eyes and white eyebrows. The ears are large, the tail bushy and flecked with brown and white hairs, and the forelimbs have sharp claws. **Behaviour** Active by day, numbats sleep in nests of shredded bark, leaves or other soft plant material carried in the mouth and placed in hollow logs or in a spherical nest chamber about 250 mm diameter at the end of a shallow, sloping burrow 1–2 m long. Some burrows are used as bolt holes and others for overnight shelter and rearing young. Numbats are solitary with home ranges of 25–50 ha, although males roam over large areas in the breeding season searching for females. They have a brisk quadrupedal gait and frequently sniff the air while standing on the hind legs. Numbats locate termites in logs and shallow galleries by scent, digging into the runways leading from the nest. They cannot breach hard termite mounds. **Development** Females are sexually mature at about 11 months, males at 2 years. They breed from January to March, and up to 4 young are born some 14 days after mating. Newborn attach to the 4 exposed teats on the mother's abdomen, clinging to hairs surrounding them and being dragged around until furred in late July. They are left in the nest while the mother forages until early September when they begin to emerge. They are weaned by late October and disperse by mid-December to establish their own home ranges, sometimes more than 15 km away. **Diet** Termites, consuming 15–20,000 per day. **Habitat** Eucalypt forests dominated by wandoo or jarrah. **Traces** Small conical scratchings around termite mounds. Scats are smooth dark cylinders with rounded ends, to 10 mm across and 10–20 mm long, deposited on fallen logs and at feeding sites. **Status** Vulnerable. Threatened by fox and cat predation and land clearing.

WEIGHT **280–715 g** HEAD–BODY **200–275 mm** TAIL **160–210 mm**

FAMILY **PERAMELIDAE** SPECIES *Isoodon auratus*

GOLDEN BANDICOOT

The size of a large rat, this marsupial was widespread throughout the central and north-western deserts until the 1930s. It is golden-brown pencilled with black, and has stiff golden and black hairs over its head and body hiding softer, greyish underfur. It has a long, pointed, somewhat flattened head, compact body and a short, pointed tail. The forelimbs have long, curved claws for digging. The long hind feet have a large, strongly-clawed fourth toe, and a double claw on the fused second and third toes for grooming. **Behaviour** They are mainly nocturnal with peaks of activity around dawn and dusk. They forage in open areas, digging small holes with the forefeet looking for food and covering up to 10 ha in a night. They are generally solitary and live within a home range of up to 36 ha. These ranges overlap extensively with other individuals. They shelter by day in simple grass nests on the ground concealed beneath vegetation or under rocks. They also shelter in caves and in large spinifex clumps and hummock grasses. They have numerous shelter sites within their home range and use different ones on consecutive days. **Development** Breeding appears to takes place throughout the year with a peak after substantial rain. Females have 8 teats in a rear-opening pouch, and usually raise 2–3 young, born about 2 weeks after mating. The young suckle in the mother's pouch for up to 8 weeks, and are then left in the nest while she forages. **Diet** Beetles, termites, ants, spiders, insect larvae, other arthropods, small reptiles, turtle eggs, rodents, roots and tubers. **Habitat** Sand dunes and sandplains with spinifex and hummock grasslands, eucalypt woodlands, heaths, coastal thickets, and rainforest margins. **Traces** Similar to the eastern barred bandicoot (page 55). **Status** Vulnerable. Threatened by land clearing, grazing and predation by introduced feral animals.

WEIGHT **250–670 g** HEAD–BODY **190–300 mm** TAIL **84–121 mm**

FAMILY **PERAMELIDAE** SPECIES *Isoodon macrourus*

NORTHERN BROWN BANDICOOT

Often found in suburban gardens, this rabbit-sized marsupial is dark brown above flecked with light brown, and pale grey to white below. The elongated head has relatively small pointed ears and small eyes. The tail is short and pointed. The forelimbs have strong curved claws on long feet. The hind feet are long with a very large, strongly-clawed fourth toe; the second and third toes are fused with a double claw used for grooming. Males are larger than females. **Behaviour** Northern brown bandicoots are nocturnal, running with an agile galloping gait and walking quadrupedally. They locate food by smell, digging conical holes with the forefeet and probing with the snout. Males are solitary and aggressive, marking areas with a scent gland behind the ear and living in home ranges of 5–6 ha. Females occupy home ranges of 1–2 ha and are followed persistently by males in the breeding season. They sleep in concealed nests of ground litter piled over shallow depressions. The nests have a hollow centre, loose entrance and exit, with soil kicked over the top for waterproofing. They use several nests and sometimes shelter in hollow logs, abandoned rabbit dens and grass tussocks. **Development** Sexually mature at 3–4 months, they live to 3 years, breeding throughout the year in Queensland, from August to March in central NSW, and August to April in the NT. Females rear several litters of 1–7 (usually 2–4) young per year, born only 12.5 days after mating. They have 8 teats in a rear-opening pouch to which the young attach for about 50 days, thereafter remaining in the nest until weaned at 2 months. **Diet** Insects, larvae, worms, berries, grass seeds and plant roots. **Habitat** Grasslands, woodlands, open forests and some closed forests, preferring areas with low ground cover including tall grass and dense shrubbery. In the NT wet season they roam in open grasslands. **Traces** Similar to the eastern barred bandicoot (page 55). **Status** Low risk.

WEIGHT **0.5–3.1 kg** HEAD–BODY **300–470 mm** TAIL **80–215 mm**

FAMILY **PERAMELIDAE** SPECIES *Isoodon obesulus*

SOUTHERN BROWN BANDICOOT

The southern brown bandicoot is a solidly-built, rabbit-sized marsupial with soft underfur and coarse, bristly guard hairs. The back is brownish-grey flecked with yellow-brown and the belly creamy-white. The tail is short with a pointed tip; the head is long and tapering with a naked nose, small rounded ears and small eyes. The short forelimbs have strong curved claws on elongated feet. The hind feet are long with a very large, strongly-clawed fourth toe; the second and third toes are fused with a double claw used for grooming. **Behaviour** Southern brown bandicoots are active mainly at night, but are occasionally seen during the day. They have a moderately fast, agile galloping gait, and walk quadrupedally, foraging on the ground for insects and worms, digging conical pits with the forefeet and probing with the snout. They sleep in an oval nest of leaf litter, grass, twigs and soil, usually built in a shallow, hemispherical ground depression up to 750 mm long. Nests are sometimes found in decaying logs, rabbit warrens, or on a raised soil platform in wet sites. Generally solitary and aggressive, their survival depends on the establishment of an adequate home range of up to 7 ha for males, and about 2 ha for females. Males range widely over areas inhabited by several females, only interacting to mate. Squeaky grunts are made during aggressive encounters. **Development** Sexually mature at 3–4 months, females begin breeding in the season following their birth, and may live to 3 years or more. The breeding season extends from late winter to the end of summer, coinciding with peaks in food abundance, and may last all year in lowland heaths, with 2–3 litters being produced each year. Up to 6 young are born 12.5 days after mating and attach to some of the 8 teats in the mother's rear-opening pouch. They are independent at 60–70 days, and the mother may give birth again as soon as the pouch is vacated. **Diet** Fungi, seeds and invertebrates including ants, beetles, larvae and worms. **Habitat** Dry sclerophyll forests and woodlands, grasslands, heaths, shrubland and regenerating areas with good ground cover; up to 1000 m. **Traces** Similar to the eastern barred bandicoot (page 55). **Status** Endangered. Threatened by land clearing and predation by feral animals.

WEIGHT **400–1600 g** HEAD–BODY **275–360 mm** TAIL **90–140 mm**

FAMILY **PERAMELIDAE** SPECIES *Perameles bougainville*

WESTERN BARRED BANDICOOT

This delicate, lightly-built, rat-sized marsupial was, prior to European settlement, widespread over much of arid southern Australia. Now is it found only on the Bernier and Dorre Islands in Shark Bay, WA. The smallest bandicoot, it has long, soft fur, mottled light brownish-grey to brownish-black above and white below, with 2–3 alternating pale and darker bars across its hindquarters. It has white feet, large erect ears and a long, pointed snout. The feet are elongated with the second and third toes of the hind feet being fused with a double claw for grooming, while the fourth toe is long and strongly clawed. **Behaviour** Solitary and nocturnal, they shelter by day in nests of grasses and other vegetation concealed in a small depression dug among litter and beneath low shrubs. They emerge at dusk to hunt and dig for food, returning to the nest at dawn in winter and several hours earlier in summer. They stay within overlapping home ranges of some 2–14 ha for males and 1–6 ha for females, with an exclusive core area, moving on average less than 400 m overnight. Most individuals use the same nest repeatedly, while others use a different nest each night. **Development** Most young are born in the wetter winter and spring months, although when conditions are favourable they breed continuously, producing up to 4 litters in a year. Females have 8 teats in their rear-opening pouch and produce 1–3 young per litter. Young are born 12.5 days after mating, spend about 60 days in the pouch, and are weaned 1–2 weeks later. They have a lifespan of up to 4 years. **Diet** Insects, spiders and other arthropods, mice, small reptiles, seeds, berries, roots and herbs. **Habitat** Scrub, saltbush and bluebush plains. **Traces** Similar to the eastern barred bandicoot (page 55). **Status** Endangered. Threatened by predation by feral animals and habitat degradation.

WEIGHT **172–280 g** HEAD–BODY **171–236 mm** TAIL **60–102 mm**

FAMILY **PERAMELIDAE** SPECIES *Perameles gunnii*

EASTERN BARRED BANDICOOT

The eastern barred bandicoot is a slender, graceful, rabbit-sized marsupial conspicuously marked with 3–4 pale bars on each side of the rump. It has soft, dense fur, mottled greyish-fawn above, paler on the flanks and pale grey to white below, with prominent coarse guard hairs. The short pointed tail is white above with a dark base. The elongated head has large pointed ears and small eyes. The short forelimbs have strong, curved claws on long feet. The hind feet are long with a very large, strongly-clawed fourth toe, while the second and third toes are fused with a double claw used for grooming. Males are larger than females. **Behaviour** Strictly nocturnal, they forage by night, digging for small invertebrates with the forefeet and probing with the snout, detecting food by sound and smell. They are solitary, sleeping by day in simple domed nests of loose grass and leaves over shallow depressions, concealed in dense undergrowth. Several nests and rabbit burrows are used for temporary shelter at night. They live in overlapping home ranges averaging 26.2 ha for males and 3.2 ha for females. When alarmed they stand upright sniffing the air, and run with an agile galloping or bounding gait, leaping up to 1 m. Vocalisations include snuffles, squeaks, hisses and grunts. **Development** Sexually mature at 4–6 months, they may live to 3 years or more, breeding from winter to early autumn with a peak from July to November. Females can raise 3–4 litters of 1–5 (usually 2–3) young per year. Born only 12.5 days after mating, the young attach to some of the 8 teats in the mother's rear-opening pouch, which they vacate after 48–55 days. They are then left in the nest while the mother forages until they are weaned at 70–80 days. Juveniles disperse at 3–5 months. **Diet** Earthworms, insects, larvae, beetles, moths and other invertebrates; seeds, roots, grass, fungi and some fruits. **Habitat** Open habitats including woodlands, open forests with a grassy understorey and grasslands, preferring deep soils and high rainfall. **Traces** Conical diggings about 100 mm wide and 150 mm deep, often with scats nearby. Scats are about 10 mm across and up to about 30 mm long, brown to black, friable with a smooth surface, containing insect remains and soil particles. **Status** Vulnerable. Threatened by predation by feral animals and habitat destruction.

WEIGHT **500–1450 g** HEAD–BODY **270–400 mm** TAIL **70–110 mm**

FAMILY **PERAMELIDAE** SPECIES *Perameles nasuta*

LONG-NOSED BANDICOOT

A common visitor to suburban gardens, the long-nosed bandicoot is a compact, rabbit-sized animal with greyish-brown fur flecked with dark brown on its back, a creamy-white belly and white feet. It has a short pointed tail, a long head with a slender muzzle and relatively long pointed ears. Its short forelimbs are designed for digging, with long toes and strong, curved claws. The second and third toes of the hind feet are fused together with a double claw used for grooming, while the fourth toe is longer and more powerful with a strong claw. **Behaviour** Nocturnal, they sleep in well-concealed nests of dry grass and plant matter piled over shallow depressions, with a hollow centre but without a visible entrance. Soil is sometimes kicked over the top for camouflage and waterproofing. Rabbit burrows are occasionally used. They forage on the ground for worms and insects, digging conical holes with the forefeet and probing with the snout, often making shrill, grunt-like squeaks. They have a quadrupedal walk and can move quite fast with a galloping gait. Adults are solitary and aggressively territorial; both sexes occupy separate home ranges, often using different nests over several nights. Males establish dominance hierarchies and follow females in the breeding season, waiting for a chance to mate. **Development** Sexually mature at 4–5 months, they may live for 2–3 years and breed at any time of year with a lull from late autumn to mid-winter. Several litters of 1–5 (usually 2–3) young are reared per year, born only 12.5 days after mating. Newborn are about 13 mm long and weigh about 0.25 g. They attach to the 8 teats in the mother's rear-opening pouch, opening their eyes at 45–50 days. They then leave the pouch and remain in the nest until weaned at about 2 months. **Diet** Invertebrates such as worms, insects and larvae; eggs, fungi, seeds and plant roots. **Habitat** Damp sites in rainforests, wet and dry sclerophyll forests, woodlands, heaths and grasslands. **Traces** Similar to the eastern barred bandicoot (page 55). **Status** Low risk.

WEIGHT **850–1100 g** HEAD–BODY **310–430 mm** TAIL **120–165 mm**

FAMILY **PERAMELIDAE** SPECIES *Macrotis lagotis*

BILBY

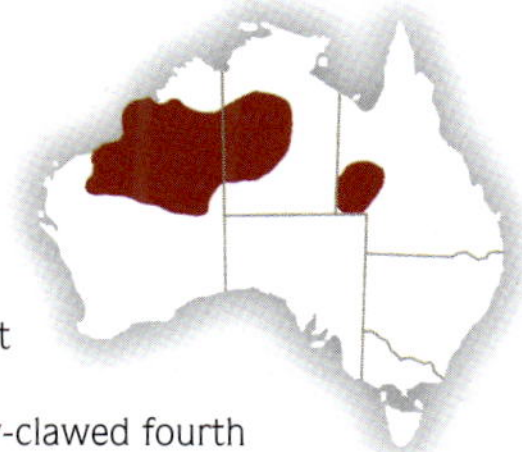

A delicately-built rabbit-sized marsupial of the desert regions, the bilby has long, soft, silky fur, bluish-grey on the back, sandy-brown on the flanks and the end of the muzzle, whitish below and on the tip of the tail. The rest of the tail is black with a crest of hairs. The narrow head has large rabbit-like ears, a long pointed snout and a long slender tongue. The short forelimbs have no first digit, the others are long with strong curved claws. The hind feet are long with a very large, strongly-clawed fourth toe; the second and third toes are fused with a double claw. Males are much larger than females. **Behaviour** Strictly nocturnal, bilbies are active mainly after midnight, sleeping in burrows up to 2 m deep and 3 m long, spiralling steeply down, often with side tunnels and several entrances. The burrows are built in termite mounds, grass tussocks or below shrubs. Adult males have a number of exclusive burrows scent-marked by their anal glands, sometimes shared with one or more females and their young. A rigid dominance hierarchy is established among males without fighting, while females have a less rigid hierarchy and utter sharp growls in aggressive encounters. They have home ranges up to 14 ha, shifting according to the food supply. Their vision is poor, but they have acute hearing and sniff out insects and fungi, digging them up with their forefeet. Bilbies often stand upright sniffing the air, and run with a cantering gait holding their tail stiffly aloft, waving it like a flag. **Development** Bilbies breed mainly from March to May, mating in the burrow and giving birth to 1–3 young after a pregnancy of 21 days. Females have 8 teats in a rear-opening pouch. Newborn attach firmly to the teats, vacate the pouch at 70–80 days, and are weaned some 14 days later. **Diet** Termites, ants, larvae, seeds, bulbs, fruit and fungi. They can survive without drinking. **Habitat** Arid and semi-arid shrublands, spinifex and tussock grasslands. **Traces** Conical diggings to 100 mm deep. Scats are often red-brown, smooth cylindrical pellets about 10 mm across and up to 25 mm long, deposited near diggings. **Status** Vulnerable. Threatened by predation by feral animals and habitat destruction.

WEIGHT **800–2500 g** HEAD–BODY **290–550 mm** TAIL **200–290 mm**

FAMILY **PERORYCTIDAE** SPECIES *Echymipera rufescens*

RUFOUS SPINY BANDICOOT

Found also in the rainforests of New Guinea, the rufous spiny bandicoot is a rabbit-sized marsupial with coarse, stiff fur; rufous-brown flecked with black and grey above, becoming blacker on the shoulders and head; straw-coloured to white below. The tail is short, black and almost naked. The head is long and narrow with small eyes, small rounded ears and a long tapering snout with a naked tip. The short forelimbs have strong curved claws on elongated feet. The hind feet are long with a very large, strongly-clawed fourth toe; the second and third toes are fused with a double claw for grooming. Males are larger than females. **Behaviour** Little studied, rufous spiny bandicoots are nocturnal and strictly terrestrial, moving with a slow hop and a galloping run. They live alone and are thought to be territorial, sheltering by day in a shallow burrow or in hollow logs and grass tussocks. They use their forefeet to dig conical holes searching for food, and investigate the hole with the snout, lapping insects up with their long, thin tongue. **Development** Little is known about their development. They have a lifespan of about 3 years and probably breed in February. Females have a well-developed, rear-opening pouch with 4 teats. **Diet** Insects, possibly small rodents, fruit and soft tubers. **Habitat** Tropical rainforests below 1200 m, vine forests, open forests, woodlands and adjoining heath **Traces** Conical diggings at feeding sites. **Status** Low risk.

WEIGHT **450–2000 g** HEAD–BODY **300–400 mm** TAIL **75–100 mm**

FAMILY **PHASCOLARCTIDAE** SPECIES *Phascolarctos cinereus*

KOALA

The koala is a short, stocky, largely arboreal marsupial with fine woolly fur, grey with brown and white patches and a white to yellowish chest and belly. It has a broad, flat head with large hairy ears, a naked flattened nose and small eyes. Although the koala's tail is no more than a stump, it is well adapted to an arboreal life with long limbs and strongly-clawed feet, enabling it to climb trees by digging its claws into the trunk. Powerful, opposing, thumb-like first digits on the hind feet can grip against the other toes. The forefeet have an unusual split-hand arrangement with the first 2 digits opposing the other 3, giving it a vice-like grip on branches. For grooming it has a double claw on the fused second and third toes of the hind feet. In subtropical and tropical districts koalas are smaller with shorter fur. Males are larger than females. **Behaviour** Solitary and nocturnal, koalas spend much of their time resting or feeding in the upper branches of gum trees. They climb slowly, grasping the trunk with the forefeet and pulling the hind limbs up, and can jump up to 2 m between branches. On the ground they walk on all fours, run with a bounding gait, and can swim if necessary. Koalas spend most of their lives within a home range, averaging about 1.7 ha for males and about 1.2 ha for females. Home ranges often overlap and males roam around, establishing dominance hierarchies, communicating their status by bellowing and marking trees with secretions from their chest gland. Juveniles disperse at about 2 years of age. Young females sometimes settle near their mother, but young males usually become nomadic for 2–3 years.

Development Koalas are sexually mature at the end of their second year, but males are unable to compete for females until they are 4–5 years old. They mate in trees and usually give birth to a single young, mostly in early summer, 34–36 days after mating. Newborn attach to one of 2 teats in the mother's rear-opening pouch, which they vacate at 6–7 months, riding on the mother's back until weaned at 11–12 months. Koalas live 10–15 years, until their teeth wear out.
Diet Leaves from a range of eucalypt trees, sometimes supplemented by radiata pine, camphor laurel and coast tea tree. They can survive without drinking in winter.
Habitat Eucalypt forests and woodlands. **Traces** Claw scratches and pock marks on tree trunks, scats around the base. Scats are oval or cylindrical, mostly brown or red-brown, about 10 mm across and 20–30 mm long with a ridged surface, packed with coarse leaf fragments. **Status** Rare in SA; vulnerable in Qld, NSW and the ACT. Threatened by land clearing, logging, road kills and attacks by domestic dogs.

WEIGHT **4–13.5 kg** HEAD–BODY **648–820 mm**

Family **Vombatidae** Species *Lasiorhinus krefftii*

Northern Hairy-nosed Wombat

This bulky marsupial is one of Australia's most threatened mammals with some 100 animals surviving in one protected area of central Qld. The largest of the 3 wombats, it has a broad head and short legs with softer, longer fur than the common wombat, longer ears and a broader muzzle fringed with fine whiskers. The fur is predominantly brownish-grey or grey mottled with fawn, brown or black. The toes have strong, flattened claws for digging and the second and third toes of the hind feet are fused with a double claw for grooming. **Behaviour** Active at night, northern hairy-nosed wombats appear to be slow and clumsy, swaying from side-to-side as they walk, although they can bound along at up to 40 kph over a short distance. When resting, they curl up on their sides or sit on their haunches with their forepaws folded in front. They dig deep, cool and humid burrows with their forelegs in sandy soil along dry creek beds, bulldozing the sand clear by moving out backwards. Burrows are up to 20 m long and 3.5 m deep with up to 7 entrances and arranged in loose clusters of up to 20 burrows within a few hectares. Burrows are marked with piles of dung and splashes of urine, and linked by well-worn trails with dung piles at intervals along the trails. Up to 10 wombats occupy each burrow cluster with equal numbers of males and females. They tend to avoid encounters and live alone, although females sometimes share burrows. To minimise water loss in droughts they only emerge at the coolest time of night, graze close to the burrow and rarely move between burrow clusters. They forage in home ranges of 3–7 ha. **Development** Wombats have a lifespan of more than 20 years. Females usually breed twice every 3 years and produce a single young from November to April. The newborn attaches to one of 2 teats in the mother's rear-opening pouch and emerges 8–9 months later. It then stays in the burrow while the mother feeds, and is weaned at about 12 months. **Diet** Grasses and sedges. They can survive for long periods without drinking. **Habitat** Eucalypt woodlands, grasslands and scrub on deep sandy soil. **Traces** Scats are cube-shaped pellets about 20 mm across, light to dark brown, comprising fine green plant fragments, deposited in groups of 4–8 near burrows. Mounds of sand at burrow entrances. **Status** Critically endangered. Threatened by competition for food by introduced grazing animals and dingo predation.

Weight **27–40 kg** Head–body **870–1100 mm** Tail **30–50 mm**

FAMILY **VOMBATIDAE** SPECIES *Lasiorhinus latifrons*

SOUTHERN HAIRY-NOSED WOMBAT

Distinguished from the common wombat by its silky fur, hairy nose and slightly smaller size, this large, stocky marsupial is grey to brownish-grey above and paler below. It has a broad flattened head with narrow, pointed ears, small eyes, rodent-like incisor teeth and a split lip enabling it to pick young emerging shoots. The tail is very short and hidden by fur. The limbs are short with short toes bearing stout, flattened claws for digging. The first toe of the hind foot is very small, while the second and third toes are fused with a double claw used for grooming.
Behaviour Southern hairy-nosed wombats are nocturnal grazers, moving slowly and clumsily, but they are very alert and can run with a bounding gait at up to 40 kph. They rest on hot days in deep humid burrows, dropping their temperature to conserve energy and water. On winter days they bask in the sun. Burrow entrances are clustered to form a large central warren with smaller warrens or single burrows surrounding it, occupying an area 200–300 m across, connected by a network of tracks marked at intervals with urine, scats and scratchings. Each warren system is occupied by 5–10 wombats, but individuals tend to move between warrens. Males establish dominance hierarchies and become aggressive during the breeding season, often biting the ears, rumps and flanks. Mating takes place in the burrow.
Development Sexually mature at 3 years, they live for 14 years or more. The breeding season extends from late September to December, although they do not breed in drought conditions, and need 2–3 years of rainfall to successfully raise their young. Usually a single young is born 21 days after mating and attaches to one of 2 teats in the mother's rear-opening pouch, emerging 6–9 months later, and staying with the mother until weaned at 12 months.
Diet Grasses and herbs. They can survive for long periods without drinking.
Habitat Arid and semi-arid grasslands and open woodlands with 200–500 mm of rain annually. **Traces** Scats are cube-shaped pellets about 20 mm across; dark brown outside; comprising fine green plant fragments; deposited in groups of 4–8 near burrows, pathways and feeding sites.
Status Low risk nationally, but endangered in NSW. Threatened by land clearing, overgrazing and habitat destruction.

WEIGHT **19–32 kg** HEAD–BODY **770–940 mm** TAIL **25–60 mm**

FAMILY **VOMBATIDAE** SPECIES *Vombatus ursinus*

COMMON WOMBAT

A large, squat, burrowing marsupial, the common wombat has long, coarse, stiff fur, patchy brownish-grey or dull sandy-brown above and paler below. It has a large broad head with a flattened naked nose, small eyes and small hairy ears, and has a very short tail hidden by fur. Its short forelimbs are adapted for digging with short toes bearing stout flattened claws. The first toe of the hind foot is just a stub, while the second and third toes are fused with a double claw used for grooming. **Behaviour** Solitary and aggressive, common wombats are mostly nocturnal, although they sometimes graze and bask in the sun during the day. They shelter in burrows up to 30 m long, often with several entrances and nesting chambers with underground connecting tunnels, usually dug in slopes above creeks and gullies. Shorter burrows 2–5 m long are used for temporary refuge. Burrows are dug in a common area and some are used by different individuals. The entrances are usually 400–500 mm across with mounds of soil outside and well-worn pathways between them. If the ground is too hard they shelter in caves and hollow logs. Individuals use more than 10 burrows, visiting 1–4 each night and alternating between 3 major sleeping burrows. Although apparently slow and clumsy, wombats can bound at 40 kph over short distances. They stay within overlapping home ranges of 4–24 ha, travelling up to 3 km overnight. Individuals scent-mark and defend separate feeding sites. They have poor eyesight but acute hearing and a good sense of smell. A male trots behind a female during courtship in circles and figure of eight patterns until she allows him to copulate. **Development** Sexually mature at 2 years, wombats have a lifespan of 15 years or more. They breed year-round on the mainland, with a peak from September to November; and in winter in Tasmania. Although twins are known, a single young weighing about 0.5 g is usually born about 27 days after mating, and attaches to one of the 2 teats in the mother's rear-opening pouch. It leaves the pouch at 6–10 months and stays in the burrow, suckling until about 15 months old, remaining at heel until about 18 months old. Females mate again when weaning is complete, usually giving birth once every 2 years. **Diet** Tussock grasses, some introduced grasses, rushes and sedges; supplemented with roots, tubers, mosses and fungi. **Habitat** Wet and dry open forests, woodlands, scrubs and wet heaths to 1800 m. **Traces** Scats are cube-shaped pellets about 20 mm across, dark brown outside, comprising fine green plant fragments, usually deposited in groups of 4–8 in conspicuous places. **Status** Vulnerable in Tasmsania; rare in SA and Qld.

WEIGHT **20–39 kg** HEAD–BODY **850–1200 mm** TAIL **20–30 mm**

FAMILY **BURRAMYIDAE** SPECIES *Burramys parvus*

MOUNTAIN PYGMY-POSSUM

The only Australian mammal restricted to alpine and subalpine areas, the mountain pygmy-possum is a largely terrestrial, rat-sized marsupial. It has fine dense fur, grey-brown above, sometimes darker along the mid-back, pale grey-brown to cream below, and darker around the eyes. It has a long thin tail covered with scales and sparse, fine hair. The hind feet are capable of grasping with an opposing first toe, while the second and third toes are fused with a double claw used for grooming. **Behaviour** Active mainly at night, mountain pygmy-possums climb among rocks and shrubs looking for food, manipulating it with the hands while squatting. Seeds are stored in the nest, under bark or beneath loose soil. They sleep in nests of shredded bark, leaves or grass, carried in the curled tail and placed in rock crevices or among dense vegetation. Males and females share nests during the breeding season, but males are forced to disperse to poorer habitats on the periphery of the breeding areas when the young leave the pouch. Juvenile males are forced out after weaning, leaving females in the best habitat and 4–5 times more likely to survive over winter than males. Females occupy non-exclusive home ranges from 0.06–7.7 ha, while males range over large distances, often travelling 1.5 km overnight. They build up fat in summer and autumn, and hibernate for up to 7 months. **Development** Mountain pygmy-possums breed in their first year of life, mating from late September to mid-October, usually producing more than 4 young, 13–16 days after mating. Only those able to attach to the 4 teats in the mother's forward-opening pouch survive. At about 30 days old they are deposited in the nest, and stay there until weaned and independent, 30–35 days later. The longest living of the small terrestrial mammals, females have a lifespan of more than 11 years in captivity and 5 years in the wild. **Diet** Moths and other invertebrates, fruits and seeds. **Habitat** Rock scree and boulder-fields in heaths, sedgelands, shrublands and woodlands above 1430 m. **Traces** Transversely cracked seeds of the mountain plum pine. Scats are small and mouse-like, about 5 mm long, containing fragments of seeds, berries and insects. **Status** Endangered; restricted to Mt Hotham and Kosciuszko areas. Threatened by tourist development.

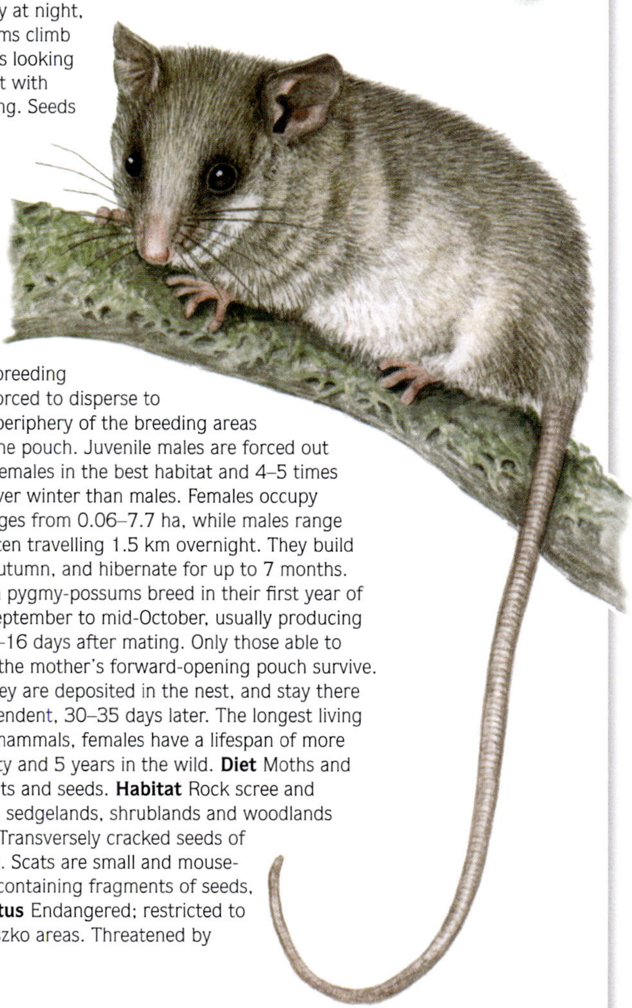

WEIGHT **30–82 g** HEAD–BODY **100–120 mm** TAIL **130–160 mm**

FAMILY **BURRAMYIDAE** SPECIES *Cercartetus caudatus*

LONG-TAILED PYGMY-POSSUM

The size of a large mouse, this arboreal marsupial is brownish-grey above with a distinct dark eye patch, and pale grey below. It has a long, thin prehensile tail with a furry, slightly thickened base. The head has a narrow, pointed muzzle with a pink nose, large eyes and large thin ears. The first toe of the hind foot is short and clawless with a bulbous tip and opposes the others for gripping branches. The second and third toes are fused with a double claw for grooming. **Behaviour** Active at night, long-tailed pygmy-possums sleep in spherical nests of leaves or fern fronds built in tree hollows, tree stumps, or in fern clumps. They are cryptic in their habits, and usually forage alone, although groups of up to 4 have been seen feeding together. They are agile climbers, using their long, prehensile tail as a fifth limb. When threatened they utter a throaty roar or a quiet defensive hiss. Lactating females may share nests with their offspring but not with other adults. Several males have been discovered sharing a nest, as have non-lactating females. Some nests contain 4–5 animals. On cool winter days when food is scarce they become torpid to conserve energy. **Development** Sexually mature at 15 months, long-tailed pygmy-possums have a lifespan of 3 years or more. They breed twice a year, with young born in January and February and from late August to early November. Up to 4 young are born per litter and the tiny newborn attach to the 4 teats in the mother's forward-opening pouch. They suckle in the pouch for about 45 days when they weigh 5–7 g. They suckle for another 35 days and live in the mother's nest until they are weaned and independent at about 3 months. **Diet** Nectar and insects. **Habitat** Tropical rainforests and open forests fringing rainforests. **Status** Low risk.

WEIGHT **20–40 g** HEAD–BODY **100–110 mm** TAIL **125–151 mm**

Family **Burramyidae** | Species *Cercartetus concinnus*

Western Pygmy-possum

The size of a large mouse, this arboreal marsupial is fawn to reddish-brown above with white belly fur. It has a conical head with large mobile ears, large eyes and very long whiskers. The tail is strongly prehensile and naked for most of its length with fine scales to give it extra grip. The tail tapers to a point from a seasonally-fattened base, and is coiled when not in use. Like other possums, the first toes of the hind feet oppose the other 3, enabling it to grasp branches, while the second and third toes are fused with a double claw for grooming. **Behaviour** Active mainly at night, western pygmy-possums move quickly through the foliage, occasionally leaping between branches, using their tail as a fifth limb. They also forage on the ground, moving on all fours and holding food in their forepaws. Solitary and cryptic, they sleep in disused bird nests or among ground litter; or construct a spherical nest of shredded bark or leaves placed in a tree hollow, stump or the crown of a grass tree. On cold or wet days they curl up into a tight ball and become torpid for periods of up to 11 days, lowering their body temperature and surviving on fat stored in the tail. **Development** Females are sexually mature at 12–15 months and breed throughout the year, producing 2–3 litters in close succession. Mating takes place soon after giving birth, although the resultant embryos remain dormant until the pouch is empty. No more than 6 young are reared. The newborn attach to the 6 teats in the mother's forward-opening pouch where they remain for about 25 days. Blind and semi-furred, the young are left in the nest until weaned at about 50 days old. **Diet** Nectar, pollen, seeds, insects and possibly small lizards. **Habitat** Dry sclerophyll forests and woodlands with a dense understorey, and semi-arid mallee scrub. **Traces** Scats are small and mouse-like, about 5 mm long, composed of very fine particles of pollen, seeds and insects. **Status** Low risk nationally, but endangered in NSW. Threatened by land clearing and predation by cats.

Weight **8–21 g** | Head–body **70–110 mm** | Tail **70–96 mm**

FAMILY **BURRAMYIDAE** SPECIES *Cercartetus lepidus*

LITTLE PYGMY-POSSUM

The smallest of the possums, this tiny, mouse-sized arboreal marsupial has soft dense fur, light grey-brown to fawn above and grey below. The head is cone-shaped with broad oval ears and large eyes. The strongly prehensile tail tapers from a seasonally fat base to a sparsely-furred, pointed tip. Little pygmy-possums can be distinguished from mice by their grasping hind feet which have an opposing first toe and a double claw on the fused second and third toes used for grooming. **Behaviour** Active at night, little pygmy-possums are fast and agile climbers, using their tail as a fifth limb. Despite this, they normally live close to the ground, foraging in the lower canopy and in thick scrub, catching and manipulating insects and small lizards with their forepaws. They are solitary and live in a drifting home range, moving to exploit flowering plants and other food resources. They sleep by day in a rough nest of grass, leaves or shredded bark in a tree hollow, crevice, tree stump, or in dense ground cover, and sometimes use abandoned bird nests. In cold conditions or when food is scarce they curl up in a tight ball and become torpid for up to 7 days, lowering their body temperature to conserve energy and surviving from fat stored in the tail. **Development** Females become sexually mature in their first year and breed throughout the year on the mainland, and from September to February in Tasmania, possibly rearing more than 1 litter per year. Usually 3–4 young are born and attach to the 4 teats in the mother's shallow, forward-opening pouch. They vacate the pouch at about 6 weeks and are left in the nest while the mother forages, or cling to the her back if she changes nests. The young are weaned at about 7 weeks. **Diet** Nectar, pollen, insects and their larvae, spiders and small lizards. **Habitat** Semi-arid mallee scrub, shrublands, wet and dry sclerophyll forests and woodlands. **Traces** Scats are difficult to find, mouse-like, 2–3 mm long, comprising very fine particles of insects and pollen. **Status** Low risk.

WEIGHT **6–13 g** HEAD–BODY **66–76 mm** TAIL **60–75 mm**

FAMILY **BURRAMYIDAE** SPECIES *Cercartetus nanus*

EASTERN PYGMY-POSSUM

The size of a large mouse, this arboreal marsupial has soft dense fur, fawn-grey to olive-brown above and light grey to white below. The rounded head has very large eyes and ears and long whiskers. The prehensile tail has a fat base in good seasons, is almost naked and tapers to a fine point. The first 2 toes of the hind feet oppose the other 3, enabling it to grasp branches, while the second and third toes are fused with a double claw for grooming. **Behaviour** Solitary and docile, eastern pygmy-possums are predominantly nocturnal, although they sometimes emerge on overcast days. They occupy overlapping home ranges of about 0.7 ha for males and 0.34 ha for females, sheltering in small nests in tree hollows and stumps, under bark, in the forks of tea trees and in dense vegetation. They also use abandoned bird nests. Several nests are used by individuals, and females may share them with their juvenile offspring. Pregnant and lactating females construct small spherical nests about 60 mm across of shredded bark lined with fresh leaves. They are largely arboreal, moving at a moderate pace among the foliage, sometimes suspended by their tail. In cold conditions they become torpid for up to 2 weeks, conserving energy and using fat reserves stored in the tail. When provoked they utter a loud hiss. **Development** Sexually mature at 4.5–9 months, they generally breed from spring to autumn on the mainland and in late winter and spring in Tasmania, giving birth to 2–4 young, raising 2–3 litters per season on the mainland in good years. Mating takes place soon after the birth of the first litter, although the embryo remains dormant until the pouch is empty. Young are born some 30 days after mating and attach to the 4 functional teats in the mother's shallow pouch (she usually has 6 teats). They leave the pouch after 33–37 days and suckle while clinging to the mother's belly fur. They are weaned at about 60–65 days, become independent immediately, and have a lifespan of 3–5 years. **Diet** Pollen and nectar gathered with the slightly brush-tipped tongue; insects, seeds and soft fruits. **Habitat** Wet sclerophyll forests, rainforests, woodlands, heaths and shrublands; from sea level to about 1600 m. **Traces** Scats are about 5 mm long and 2 mm wide, comprising fine particles of pollen, seeds and insects. **Status** Low risk nationally, but vulnerable in NSW. Threatened by land clearing, fire regimes and predation by feral animals.

WEIGHT **15–45 g** HEAD–BODY **70–110 mm** TAIL **75–110 mm**

FAMILY **PETAURIDAE** SPECIES *Dactylopsila trivirgata*

STRIPED POSSUM

The striped possum is a shy and slightly-built arboreal marsupial with a variable pattern of black and white stripes from its nose to the base of its bushy tail, and a white Y-shaped pattern on its forehead. The belly is white, the limbs black-striped, and the tip of the pendulous, prehensile tail is often white. It has large rounded ears, a large blunt head with long, sharp, lower incisor teeth and a long tongue. The hind feet are adapted for grasping branches and have an opposing first toe, while the second and third toes are fused and have a double claw for grooming. The striped possum has a larger brain for its size than any other marsupial, and an extremely long fourth finger with a powerful curved claw. **Behaviour** Active mainly at night, striped possums spend up to 9 hours per night foraging in the trees and investigating rotten logs for insects, biting into the bark with their bayonet-like lower incisors, extracting prey with their long tongue or fourth finger. They are extremely agile and fast, leaping through the tree canopy in pursuit of rivals, and running along branches with a peculiar lizard-like gait, moving their opposite front and rear legs together. During the day they sleep in leaf-lined nests in the hollows of large trees or clumps of epiphytes, using many different den sites spread over a home range of 5–22 ha. Males are generally solitary and sleep alone. Females and juveniles often share dens. They are very noisy, making guttural shrieks when mating, snorting, rustling through the foliage, tearing at the bark, and making loud chewing and slurping sounds when feeding. When handled they emit a strong musky odour. **Development** They have a lifespan of about 5 years in the wild and mating probably occurs year round with a peak in June–July. Females usually give birth to 2 hairless young with their eyes closed. Newborn attach to the 2 teats in the mother's well-developed, forward-opening pouch. The young are carried on the mother's back after weaning. **Diet** Mostly wood-boring insects and larvae, supplemented by leaves, pollen, flowers, sap, fruits, small vertebrates and honey. **Habitat** Lowland tropical rainforests, gallery forests, vine forests and adjoining woodlands. **Traces** Feeding scars on tree trunks. **Status** Low risk.

WEIGHT **240–569 g** HEAD–BODY **240–280 mm** TAIL **310–394 mm**

Family **Petauridae** Species *Gymnobelideus leadbeateri*

Leadbeater's Possum

Once presumed extinct, this arboreal marsupial was rediscovered in 1961. It has soft fur, grey to greyish-brown above with a prominent dark brown stripe along the back, and paler below. The ears are thin, large and rounded with a small white patch at the base. Arboreal adaptations include a thick prehensile tail, broader and more bushy at the tip, a clawless opposable first toe on the hind foot, large striated footpads for extra grip and retractable claws. The second and third toes are partly joined with a double claw for grooming. **Behaviour** Leadbeater's possums are shy and secretive. They emerge from their nest at dusk and move rapidly through the canopy, making spectacular leaps between trees and climbing smooth tree trunks with ease. Up to three quarters of their time is spent in a communal nest more than 250 mm across, made of shredded bark, built in the hollow centre of a large tree with a narrow access hole. Members of a colony often move between tree hollows, using nests in 2–3 trees around 50 m apart and defend a territory of about 1–3 ha against intruders from adjacent colonies. Each colony usually comprises up to 8 animals, with a monogamous breeding pair, their offspring, and sometimes an extra male. Females are more aggressive than males and evict their daughters at 10 months, forcing them to establish a new colony or join one which has lost its resident breeding female. Males disperse at about 15 months to join a neighbouring colony or become part of a bachelor group until mating opportunities arise. Colony members groom each other frequently and recognise each other by smell. Vocalisations include hissing and chattering. **Development** Females are sexually mature at 10 months, males at 15 months. They breed in all months except January and February with birth peaks in May–June and October–November. Females have 4 teats in a rear-opening pouch and give birth to 1–2 young about 20 days after mating. They attach to the teats and remain in the pouch for 80–93 days, emerging from the nest at about 110 days to forage alone. **Diet** Beetles, spiders, crickets and other insects extracted from shedding bark; honeydew, and exudates licked from plants. **Habitat** Tall, open, montane forests, mainly mountain ash, with a dense understorey and large old or dead trees. **Traces** Scats are about 3 mm across and 20 mm long, friable, containing fine powdery particles, with a strong, sweet odour when fresh. **Status** Endangered. Threatened by logging.

Weight **100–160 g** Head–body **150–170 mm** Tail **145–180 mm**

FAMILY **PETAURIDAE** SPECIES *Petaurus australis*

YELLOW-BELLIED GLIDER

This rabbit-sized gliding possum is dark to pale grey-brown above, off-white to yellow-orange below, with dark stripes along the mid-back and thigh, and dark limbs. The ears are large, pointed and almost naked and a square gliding membrane joins the wrists and ankles. Arboreal adaptations include a long, broad and bushy prehensile tail and grasping feet. The first 2 digits of the forefeet oppose the other 3, and the hind feet have an opposing first toe. The second and third toes of the hind feet are fused with a split claw used for grooming. **Behaviour** Active mainly at night, yellow-bellied gliders are fast and very agile, running beneath branches and gliding up to 120 m between trees. They travel more than 2 km from the den to forage, occasionally feeding in small groups in the same tree. Their social behaviour is among the most complex of any marsupial studied, and includes a wide range of vocalisations from loud gurgling shrieks to growls, whoos and rattling chatter. They pair for life and live and move in family groups of 3–6 individuals, sometimes with more than one breeding female. Groups use dens in large tree hollows, building substantial spherical nests of eucalypt leaves carried in the coiled tail. Each group uses several hollows within an exclusive home range of 30–65 ha. Adult males scent-mark family members with secretions from glands on the forehead, chest and base of the tail. Males are heavier than females.

Development Mating often takes place while clinging to the underside of a branch. Births occur at any time of year with peaks from August to September in the south and May to September in the north. Usually a single young is born and attaches to one of 2 teats in the forward-opening pouch, where it stays for 90–100 days, remaining in the nest for another 40–50 days, becoming independent at 6–8 months. Both parents raise the young, and juveniles remain with their parents until sexually mature at 18–24 months, when they disperse. They have a lifespan of 14 years or more. **Diet** Nectar, pollen, insects, manna, honeydew and eucalypt sap obtained by biting into the bark. **Habitat** Tall, mature eucalypt forests and woodlands, often in mountainous areas. **Traces** Deep, V-shaped slashes in eucalypt bark; oval pellets of chewed bark on the ground. Scats are shiny dark brown, about 5 mm across and 15 mm long, containing fine powdery material and sometimes insect remains. **Status** Vulnerable. Threatened by logging

WEIGHT **450–700 g** HEAD–BODY **270–320 mm** TAIL **420–480 mm**

Family **Petauridae** Species *Petaurus breviceps*

Sugar Glider

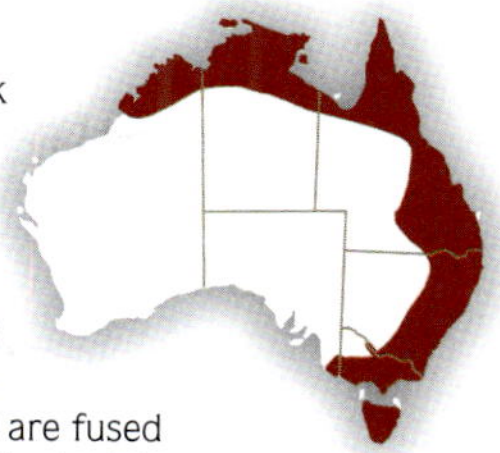

This small gliding possum is light grey above with a distinct dark stripe along its forehead and back, and dark patches behind its large pointed ears. It is creamy-white to pale grey below with short body fur and a bushy prehensile tail, sometimes white-tipped. A square gliding membrane joins the fifth finger and the first toe. The feet are able to grasp branches: the first 2 digits of the forefeet oppose the other 3, and the hind feet have an opposing first toe. The second and third toes of the hind feet are fused and have a split claw used for grooming. **Behaviour** Active mainly at night, sugar gliders are shy, fast, agile climbers and can glide more than 50 m. They build communal spherical or bowl-shaped nests of eucalypt leaves in tree hollows, shared by up to 7 adults and their young of the season. A typical social group comprises 3–4 adult females and 2–3 adult males. The dominant male scent-marks other group members and the group's territorial boundaries with saliva and secretions from glands on his forehead, chest and at the base of his tail. Visitors without the appropriate scent marking are violently expelled. Territories of around 2.5 ha and favourite food trees are defended against intruders. Young males are forced to disperse, but vacant areas are scarce, and few survive. Young females sometimes remain to replace females lost to the group. In very cold, wet weather they huddle together in the nest and become torpid for short periods. Vocalisations include a loud bark or 'yip' when alarmed and a rolling guttural anger call. **Development** Females become sexually mature at 8–15 months, males at 12 months. They have a life expectancy of 4–5 years in the wild and breed from June to November (year-round in the NT), rearing up to 2 litters per season. Twins are usually born 16 days after mating and attach to 2 of 4 teats in the forward-opening pouch for 40 days, emerging at 60–70 days. Young remain in the nest for another 50 days and forage with their mother until 7–10 months old. **Diet** Acacia gum and eucalypt sap licked from gashes in tree trunks made with their lower incisor teeth; nectar, pollen, manna, honeydew, beetles, moths, spiders and other invertebrates; occasionally small birds and mammals. **Habitat** Wet and dry eucalypt forests and woodlands, usually with wattles; from sea level to over 1200 m. **Traces** Similar to the squirrel glider (page 72). **Status** Low risk nationally, but rare in SA.

Weight **95–170 g** Head–body **150–210 mm** Tail **165–210 mm**

FAMILY **PETAURIDAE** SPECIES *Petaurus norfolcensis*

SQUIRREL GLIDER

Larger, but otherwise similar to the sugar glider, this species is light grey above with a dark stripe along the forehead and back. It is white to creamy-white below with white markings on the cheeks and behind the ears. The prehensile tail is broad, long, bushy and soft with a dark tip. It has large protruding eyes, large ears, a pointed snout and a rectangular gliding membrane between the hands and feet. The feet are grasping: the first 2 toes of the forefeet oppose the other 3 and the hind feet have an opposing first toe. A double claw on the fused second and third toes of the hind feet is used for grooming. **Behaviour** Active mainly at night, squirrel gliders are agile climbers and can glide more than 50 m between trees. They forage in the upper and lower forest canopy, gleaning insects from the foliage and licking sap from gashes made with their sharp incisor teeth in the bark of eucalypt and acacia trees. They build bowl-shaped, leaf-lined nests in tree hollows, utilising a number of hollows and travelling up to 1 km from preferred hollows to foraging sites. They nest alone, in pairs or in family groups of up to 8 individuals comprising one mature male, generally 2 females, and juveniles of the season. Males have well-developed scent glands on their foreheads which they use to mark their territory, and family groups occupy home ranges of around 3 ha, depending on availability of resources. Vocalisations include a deep, throaty, gurgling chatter and soft grunts. **Development** Sexually mature at about 12 months, squirrel gliders have a lifespan of up to 15 years. They give birth throughout the year with a peak from May to December, often producing twins, and can raise 2 litters in a year. The young are born 20 days after mating and attach to 2 of the 4 teats in the mother's forward-opening pouch. They vacate the pouch at about 70 days and are left in the nest for another 30 days before accompanying the mother on foraging trips until weaned at about 6 months. Juveniles remain in the family home range until they are 10–12 months old. **Diet** Nectar, pollen, flowers, acacia gum, eucalypt sap, caterpillars and other insects. **Habitat** Dry sclerophyll forests and woodlands with abundant tree hollows. **Traces** Horizontal and vertical gashes on tree trunks. Scats are found at the base of trees. They are roughly formed cylinders less than 5 mm across and 10–15 mm long, with a dark shiny surface, usually friable, containing insect fragments and powdery particles. **Status** Low risk nationally, but vulnerable in NSW. Threatened by land clearing and predation by cats and foxes.

WEIGHT **170–300 g** HEAD-BODY **170–240 mm** TAIL **220–300 mm**

FAMILY **PSEUDOCHEIRIDAE** SPECIES *Hemibelideus lemuroides*

LEMUROID RINGTAIL POSSUM

A predominantly arboreal marsupial usually found high in the forest canopy, the lemuroid ringtail possum can be spotted by its bright yellow eyeshine. It has soft fur, uniformly dark charcoal-grey above, sometimes brownish on the shoulders and lighter below with a yellowish tinge. A rare form is creamy-white with a tinge of orange on the shoulders. The face is short with a roman nose and small ears. The long prehensile tail is bushy, slightly tapered with a short, naked, finger-like tip, coiled when not used. To help them grip branches the first 2 fingers of the hands oppose the other 3 and the hind feet have an opposing first toe. Long, curved, sharp nails on the digits act as grappling hooks. The second and third toes are fused together with a double claw used for grooming. **Behaviour** Strictly nocturnal, lemuroid ringtail possums sleep in dens in tree hollows, often sharing with other members of the same family. They have a strong, long-lasting pair bond, are very gregarious and often seen at night in pairs or family groups, sometimes feeding in a tree with other groups or individuals. They are noisy animals, frequently leaping 2–3 m through the canopy with their legs outstretched like a glider and crashing into the foliage. They forage in the canopy on average 16 m high. They occasionally emit loud screams and wails and produce a musky-smelling, sticky, creamy-coloured secretion from their anal glands when handled. On the ground they move on all fours. **Development** Lemuroid ringtail possums have a lifespan of 4–6 years and breed from June to July, producing a single young in early August that attaches to one of 2 teats in the mother's well-developed, forward-opening pouch. The young leave the pouch at 6–7 weeks old and ride on their mother's back for another 6 months. If separated the young emit a high-pitched squeaking distress signal. Both parents care for the young until they become independent at about 9 months old. **Diet** Mainly leaves, supplemented by flowers and fruits. **Habitat** Tropical rainforests above 450 m. **Traces** Brilliant white-yellow eyeshine at night. Scats are dark oval to cylindrical pellets about 5 mm across and 10 mm long, packed with fine plant material. **Status** Low risk.

WEIGHT **750–1300 g** HEAD–BODY **300–380 mm** TAIL **300–375 mm**

FAMILY **PSEUDOCHEIRIDAE** SPECIES *Petauroides volans*

GREATER GLIDER

The largest of the gliding possums, the greater glider is whitish below and dark grey, dusky-brown, cream or greyish-white above. It has large ears with a furry fringe, a short snout and a long, bushy, pendulous and weakly-prehensile tail. A square membrane joins the flanks, elbows and ankles and acts as a wing when the limbs are extended. The feet are designed to grip onto branches: the first 2 toes of the forefeet oppose the other 3 and the hind feet have an opposing first toe. A long, split claw on the fused second and third toes of the hind feet is used for grooming. **Behaviour** Active at night, greater gliders emerge at dusk from dens high up in tree hollows, sometimes lined with bark strips or leaves carried in the curled tail, and follow established routes to feeding sites. They are generally solitary, although during the mating season pairs often share dens until the young leave the pouch. Males and females occupy individual home ranges of about 1.3–4 ha, defined by scent-marking trees with anal gland secretions and urine and use up to 18 den sites. Male home ranges rarely overlap, but female home ranges often overlap those of males or other females, and in good habitats males may have exclusive access to several females. They are agile climbers and can glide more than 100 m, extending the limbs, bringing the paws under the chin and steering with the tail and gliding membrane. On the ground they have a clumsy loping gait. Females returning to the den after foraging communicate with their nestlings with high-pitched twittering sounds. **Development** They are sexually mature at about 21 months and have a lifespan of up to 15 years. Mating takes place from March to June. A single young is born and attaches to one of 2 teats in the mother's forward-opening pouch. The infant leaves the pouch at 3–4 months and for the next 3 months is left in the den or rides on the mother's back while she climbs, but not while gliding. Young are independent at 8–10 months. **Diet** Eucalypt leaves and buds, and occasionally mistletoe. They rarely need to drink. **Habitat** Wet and dry sclerophyll forests and tall woodlands with old hollow trees; from sea-level to above 1200 m. **Traces** Bright eyeshine at night and scratches made on tree trunks when landing. Scats are found under den trees. They are pellets, 5–10 mm across, red-brown to brown, comprising fine leaf particles. **Status** Low risk.

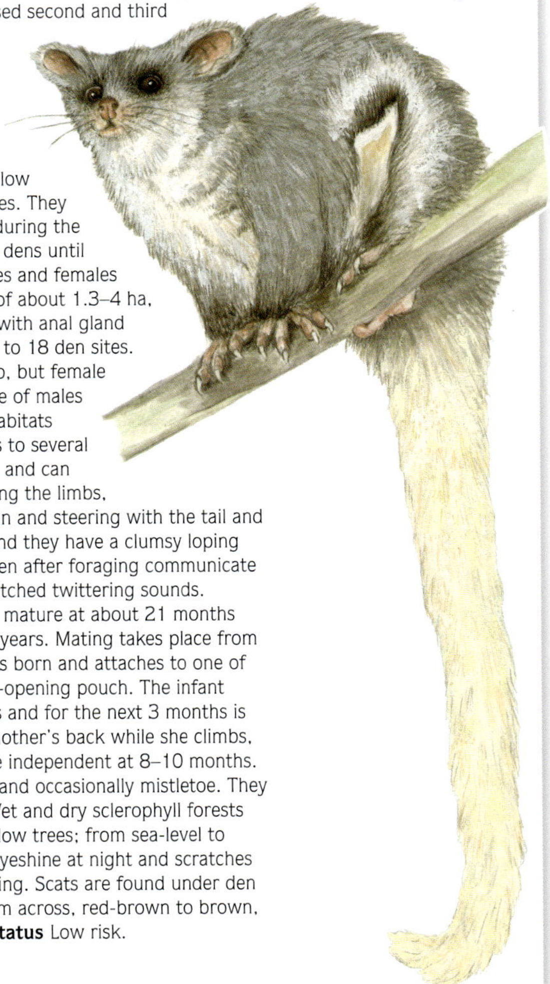

WEIGHT **900–1800 g** HEAD–BODY **350–450 mm** TAIL **450–600 mm**

FAMILY **PSEUDOCHEIRIDAE** SPECIES *Petropseudes dahli*

ROCK RINGTAIL POSSUM

A rabbit-sized marsupial, the rock ringtail possum has long woolly fur, grey to brownish-grey above with a dark stripe from forehead to rump, white below with white patches around the eye and under the ear. The short, prehensile tail is thick at the base and tapers to a thin, naked tip. The head is pointed with a naked nose and small rounded ears. The legs are short with short claws on hands and feet. The first 2 fingers of the forefeet oppose the other 3. The hind feet have an opposing first toe; the second and third are fused with a double claw for grooming. **Behaviour** Active at night, rock ringtail possums emerge soon after dark and move in family groups down the rocks to feed in nearby trees and shrubs. They sleep by day on well-protected rock ledges, using clefts and fissures as cool, secure den sites. They live in family groups, maintain long-term consorts, touch each other frequently and share dens within non-overlapping home ranges. Adults mark rocks and tree branches with secretions from their chest and anal glands to define their home range. They are shy and run for cover when caught in a spotlight, sometimes hiding their head in a rock cleft, but leaving the body exposed. They are good climbers and leapers. On the ground they walk and run on all fours, and sit on their tail when resting. Vocalisations include quiet screeches and grunts. Groups of possums have a coordinated sentinel system, alerting each other to predators. **Development** Rock ringtail possums are thought to breed year-round, and females with large young in the pouch have been seen from March to September. They give birth to a single young that attaches to one of 2 teats in the mother's forward-opening pouch. Young ride on their mother's back after leaving the pouch, and spend time at heel before becoming independent. Males contribute equally to parental care, marshalling the young and watching for predators. **Diet** Leaves, flowers and fruits of trees and shrubs. **Habitat** Rocky escarpments and outcrops with deep fissures in open forest and vine forest thickets. **Traces** Bright red eyeshine; does not freeze in a spotlight. Scats are reddish-brown to black, banana-shaped, 15–25 mm long and about 5 mm across, usually deposited in groups of 3–6. **Status** Low risk.

WEIGHT **1.2–2 kg** HEAD–BODY **330–385 mm** TAIL **200–266 mm**

Family **Pseudocheiridae** | Species *Pseudochirops archeri*

Green Ringtail Possum

A rabbit-sized arboreal marsupial of dense upland rainforests, the green ringtail possum has thick soft fur, greyish-green to lime-green above with 2 silvery-yellow stripes from the shoulders to the rump on each side of the spine. It is white below with white patches around the eyes and ears. The pointed head has a pink nose, bulging eyes and small rounded ears. Arboreal adaptations include a relatively short prehensile tail with a thick base tapering to a narrow tip that is curled up when not used, and opposing fingers and toes for gripping. The first 2 digits of the forefeet oppose the other 3, while the hind feet have an opposing first toe and fused second and third toes with a double claw for grooming. **Behaviour** Green ringtail possums are active mainly at night, although if disturbed they may feed and move during the day. Most of their active time is spent high up in the canopy where they move with surprising speed and agility, running along branches and up vines, but avoiding leaping. They rarely descend to the ground, doing so only to reach another tree, moving on all fours. They sleep upright on a branch among the foliage, curled into a tight ball with the tail coiled under the chin, gripping the branch with their hind feet. The green ringtail possum is perhaps the most solitary of the ringtails, although mothers with young at heel and pairs of mating adults are occasionally seen. Adults are silent, but pouch young utter a quiet 'tssk tssk tssk' when handled. **Development** Sexually mature at about 12 months, they probably breed all year with a peak from July to November, giving birth to a single young that attaches to one of 2 teats in the mother's forward-opening pouch. Young ride on the mother's back after vacating the pouch, and then follow the mother for several weeks as she forages in the trees. **Diet** Leaves, predominantly from fig trees, and ripe figs. Soft faecal pellets of partially digested food are produced during the day and consumed immediately. **Habitat** Dense tropical rainforest above 300 m. **Status** Low risk.

Weight **670–1350 g** | Head–body **285–380 mm** | Tail **310–372 mm**

Family **Pseudocheiridae** | Species *Pseudochirulus herbertensis*

Herbert River Ringtail Possum

An arboreal marsupial of the dense tropical rainforests, the Herbert River ringtail possum has dark brown to black fur with white or cream markings on the chest, belly and upper forearm. Some adults have no white fur, while juveniles are pale fawn and have a stripe along their upper head and back. It has a long pointed face, small rounded ears, bulging eyes and a narrow, tapering, prehensile tail carried in a coil. To help it grip branches the first 2 fingers of the forefeet oppose the other 3 and the hind feet have an opposing first toe, while the second and third toes are fused with a double claw used for grooming.

Behaviour Active at night, they emerge from their dens shortly after sunset and return 50–100 minutes before dawn. They are agile, cautious climbers, making small, careful leaps through the canopy and rarely descending to the ground where they walk on all fours. They sleep by day in dens in tree hollows, fern clumps or dense vegetation, sometimes building rough spherical nests using shredded bark or other vegetation carried in the curled tail. They are generally solitary outside the breeding season and occupy overlapping home ranges of 0.5–1 ha. During the breeding season, a female on heat is accompanied by a consort male who stays close to her throughout the night, shadowing her for up to 48 days until she allows him to mate. Pairs share a nest in the breeding season and generally relocate each year. Vocalisations consist of quiet clicks, grunts and screeches. **Development** Herbert River ringtail possums are sexually mature at 12 months. They mate from March to December with a peak in May–July, usually producing one litter of twins each year. Newborn attach to the 2 developed teats in the mother's forward-opening pouch, which they vacate at 105–120 days. Young ride on the mother's back for 2 weeks, and are then left in the nest until weaned at 150–160 days. **Diet** Leaves, supplemented with flowers and fruits. **Habitat** Tropical rainforest above 350 m and occasionally tall open eucalypt forest bordering rainforests. **Traces** Bright pinkish-orange eyeshine at night. Scats are dark, about 5 mm across and 15 mm long, containing fine plant fragments. **Status** Low risk.

Weight **800–1530 g** | Head–body **300–400 mm** | Tail **290–470 mm**

FAMILY **PSEUDOCHEIRIDAE** SPECIES *Pseudocheirus peregrinus*

COMMON RINGTAIL POSSUM

Often seen in suburban gardens, this rabbit-sized arboreal marsupial has a grey back, sometimes rufous-tinged, white to bright-rufous underparts, pale patches below the ears and eyes, and rufous-tinged legs. It has a long, white-tipped prehensile tail carried in a loose coil when not in use, bare for some distance below to help it grip branches. Other arboreal adaptations include strongly clawed feet with the first 2 fingers of the forefoot and the first toe of the hind foot opposing the others. The second and third toes of the hind feet are fused with a long, split claw used for grooming. **Behaviour** Active at night, they spend most of their time foraging and grooming in trees and leaping between branches. On the ground they move on all fours and can swim well. They rest in spherical nests around 300 mm diameter with a circular entrance, built in the fork of a tree or shrub using woven twigs, leaves, fern fronds and grass. Nests are lined with shredded bark, moss or grass. Tree hollows are used, but only when available, allowing them to exploit a variety of vegetation types. They live in family groups comprising an adult male with one or occasionally 2 adult females and offspring from the previous breeding season. Families build and use a number of nests and occupy a common home range of around 0.37 ha, marked by scent secretions. Juveniles are forced to disperse and establish their own home ranges. The home ranges of females overlap, while males tend to have exclusive home ranges. They emit soft, high-pitched twittering calls and when handled they secrete a strong-smelling liquid from their anal glands. **Development** Sexually mature at about 13 months, they have a lifespan of 3–6 years. Litters vary from 1–4, but usually 2 young are born from late April to December, and 2 litters a year may be raised. Newborn attach to one of the 4 teats in the mother's forward-opening pouch for 42–49 days. They leave the pouch at about 4 months and are left in the nest for the next 2 weeks or carried on their mother's back while she forages. Fathers often help to raise the young, carrying them around and nesting with them during the day until they are weaned at 5–8 months. **Diet** Eucalypt and other leaves, flowers and fruits. Soft faecal pellets of partially digested food are produced during the day and eaten immediately. **Habitat** Forests, woodlands, rainforests, tea tree thickets and dense shrublands; to above 1200 m. **Traces** Claw marks on tree trunks; small groups of scats, chewed leaves and blossoms at the base of trees. Scats are red-brown to green, cylindrical with rounded ends, 5–8 mm across and about 15 mm long. **Status** Low risk.

WEIGHT **660–1100 g** HEAD–BODY **300–350 mm** TAIL **300–350 mm**

FAMILY **TARSIPEDIDAE** SPECIES *Tarsipes rostratus*

HONEY POSSUM

This mouse-sized arboreal marsupial has a very long snout and a long, brush-tipped tongue. It has coarse fur, grizzled greyish-brown above and cream below with 3 dark stripes along the back. The eyes are near the top of the head and the ears are rounded. The tail is long, thin, strongly prehensile and sparsely furred. The teeth are rudimentary pegs except for 2 long, pointed, lower incisors. The hands and feet have opposing first digits and large rough pads for gripping branches. The second and third toes of the hind feet are fused and used for grooming; the claws are reduced to small nails. **Behaviour** Active mainly at night, honey possums emerge at dusk to feed. They are very fast moving and agile, darting between blossoms, extracting pollen and nectar with their long, brush-tipped tongue. They run on all fours with the tail held straight out behind, and can walk below branches and climb vertically. They live in overlapping home ranges of about 0.75 ha for males and 0.15 ha for females. They do not build nests, but shelter in the hollow trunks of grass trees or abandoned bird nests, sometimes sharing den sites. On cold days they huddle together for warmth and become torpid for short periods to save energy when food is short. Females dominate males, and several males may compete to mate with a female. **Development** Sexually mature at about 6 months, with a life expectancy of 1–2 years, they breed throughout the year with peaks in early autumn, winter and spring when food is most abundant, producing 2–3 litters per year. Up to 4 young are born 21–28 days after mating. Newborn weigh about 5 mg and are the smallest mammalian young. They attach to the 4 teats in the mother's forward-opening pouch, emerging at about 65 days, when they are left in a shelter or nest while the mother forages. The young disperse after weaning at 90 days. Females carry dormant embryos which develop when the pouch is empty. **Diet** Pollen and nectar. **Habitat** Sandplain heaths and shrublands with nectar-producing plants (notably Proteaceae and Myrtaceae families). **Status** Low risk.

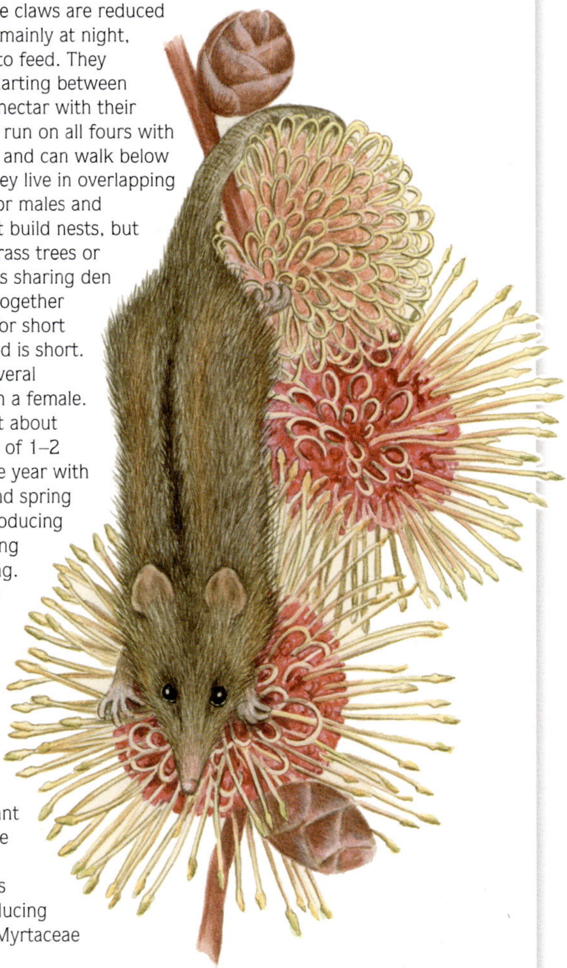

WEIGHT **7–17 g** HEAD–BODY **40–95 mm** TAIL **45–110 mm**

FAMILY **ACROBATIDAE** SPECIES *Acrobates pygmaeus*

FEATHERTAIL GLIDER

The feathertail glider is the smallest and most cryptic of the gliding possums. The size of a mouse, it has a flattened, feather-like tail with long stiff hairs on each side, used as a steering aerofoil and as an extra limb. A small, thick, gliding membrane extends from the elbows to the knees. The feet have a powerful grip with sharp claws, 2 opposing toes on the forefeet, opposing first toes on the hind feet, and finely-grooved pads enabling it to climb smooth surfaces. A double claw on the fused second and third toes of the hind feet is used for grooming. It has a brush-tipped tongue for feeding on nectar and large, forward-pointing eyes. **Behaviour** Active mainly at night, feathertail gliders are fast and agile, foraging on the ground and in trees to 40 m high, leaping through the tree canopy, gliding up to 25 m. They share bulky, spherical nests of leaves, ferns and bark strips in tree hollows or crevices, huddling together in cold or wet weather, and becoming torpid for short periods when food is scarce. Up to 20 individuals may live in the same tree hollow, although groups of 5–6 are more common, with females tending to outnumber males. Females form matrilineal groups comprising mothers and their female offspring. Young males disperse at about 6 months of age, and males appear to be solitary outside the breeding season. **Development** Females become sexually mature at about 8 months, while males are sexually mature at about 12 months. They have a maximum lifespan of 9 years. Births occur between April and February in the south, and year-round in the north. Females mate immediately after the birth of the first litter, the new embryo remaining dormant until the first litter is weaned, hence 2–3 litters may be raised in quick succession. Litter size varies from 1–4, although 2–3 is more usual. Females have 4 teats in a forward-opening pouch where the young remain for 60–65 days, staying in the nest until weaned and independent at 95–100 days. **Diet** Pollen, nectar, insects, honeydew and manna. Sap is licked from gashes made by sugar gliders and yellow-bellied gliders. **Habitat** Wet and dry sclerophyll forests and woodlands from sea level to above 1200 m. **Traces** Scats are small and mouse-like, about 5 mm long and 1 mm across, composed of very fine particles. **Status** Endangered in SA. Threatened by logging of old, hollow trees and predation by cats and foxes.

WEIGHT **10–17 g** HEAD-BODY **60–80 mm** TAIL **60–80 mm**

FAMILY **PHALANGERIDAE** SPECIES *Spilocuscus maculatus*

SPOTTED CUSCUS

An arboreal marsupial of the northern tropical rainforests, the spotted cuscus is heavily-built with a very short snout and a flat round face with large, forward-pointing, red-rimmed eyes. The dense woolly fur is creamy-white below and hides the small ears. Males may be a uniform creamy-white above or blotched with red-orange, black or grey. Females are usually unblotched and may have a black or grey saddle on the back. The long, strongly prehensile tail is naked for the terminal two-thirds with a rough undersurface for extra grip, and is coiled up tightly when not in use. The feet have strong curved claws and the first 2 toes of the forefeet oppose the other 3 enabling it to grasp branches. They have long canine teeth. **Behaviour** Active mainly at night and on cool, overcast days, the spotted cuscus is usually seen alone, resting on exposed branches in dense clumps of foliage, or sleeping on a small temporary platform of leaves and twigs pulled underneath itself, rarely returning to the same tree the following day. Males are territorial and aggressively defend an area defined by scent-marking. They climb slowly and deliberately through the canopy, sometimes leaping between trees, holding branches in a vice-like grip with the feet, using the tail as a fifth limb. They occasionally descend to the ground, moving with a bounding gait, and may travel a few hundred metres. Vocalisations include hisses, screeches, grunts and clicks. If stressed they secrete a reddish-brown substance on the bare facial skin, particularly around the eyes. **Development** Spotted cuscuses have a lifespan of 11 years or more and are sexually mature at about 8 months old. They probably breed throughout the year, usually giving birth to a single young (up to 4 may be produced) some 13 days after mating. The newborn attach to some of the 4 teats in the mother's well-developed, forward-opening pouch. They leave the pouch at 6–7 months old and are carried on the mother's back until weaned. **Diet** Leaves, fruits, insects and possibly small birds and mammals. **Habitat** Mainly rainforests from sea-level to 820 m, some open forests and mangroves. **Traces** Scats are brown cylindrical pellets up to 10 mm across and 20–30 mm long, comprising coarse plant material. **Status** Rare.

WEIGHT **1.5–4.9 kg** HEAD–BODY **350–580 mm** TAIL **315–435 mm**

FAMILY **PHALANGERIDAE** SPECIES *Phalanger intercastellanus*

SOUTHERN COMMON CUSCUS

Previously known as the grey cuscus, this species is a relatively lightly-built arboreal marsupial of the northern tropical rainforests. It has soft woolly fur, grey-brown above and off-white below, with a brown stripe from the ears to the rump. Males have a distinct yellowish chest gland and a yellow tinge on the side of the neck. The triangular head has bulging, forward-facing eyes and small, round, projecting ears. Arboreal adaptation includes a long prehensile tail, naked for two-thirds of its length, with a rough gripping undersurface, and grasping forefeet with the first 2 toes opposing the other 3. They have strong curved claws and long canine teeth. **Behaviour** Cryptic and solitary, southern common cuscuses sleep in dens in tree hollows during the day and climb slowly through the canopy at night, leaping across gaps, gripping branches with the forefeet and using the tail as a fifth limb. On the ground they move with a slow bounding gait. Males are probably territorial, using their chest gland to scent-mark areas. Vocalisations include harsh aggressive guttural screeches, grunts and clicks. **Development** The southern common cuscus probably breeds throughout the year. Females usually give birth to twins that attach to 2 of the 4 teats in the mother's well-developed, forward-opening, pouch. Infants are carried on the mother's back after vacating the pouch. Males grow throughout their life. **Diet** Leaves, fruits, flowers, buds and seeds. **Habitat** Tropical rainforests at all elevations. **Traces** Bright red eyeshine at night. **Status** Rare. Threatened by habitat loss and hunting.

WEIGHT **1.5–2.2 kg** HEAD–BODY **350–400 mm** TAIL **280–350 mm**

Family **Phalangeridae** Species *Trichosurus caninus*

Mountain Brushtail Possum

A relatively large and robust arboreal marsupial, the mountain brushtail possum is flecked steel-grey above and whitish below. It has short, rounded ears and a pink nose. Arboreal adaptations include a black, bushy, curved and tapering prehensile tail with a bare section beneath the tip, and clawless opposing first toes on the hind feet for gripping. The forefeet have 5 strongly-clawed, non-opposing fingers, while the second and third toes of the hind feet are fused with a long, split claw used for grooming. Animals in north eastern districts are dark grey to deep amber-brown above. **Behaviour** Active mainly at night, mountain brushtail possums move rapidly through the trees, often crashing through the canopy and frequently descending to the ground to feed. They walk on all-fours in a rolling gait with the hind feet turned out, and can swim if necessary. They sleep in dens in tree hollows, hollow logs, stumps or occasionally in large epiphytic ferns. Most adults use more than one den and both sexes have overlapping home ranges of about 7 ha for males and 5 ha for females. Home ranges are marked with secretions of scent glands located under the chin, on the chest, and below the base of the tail. Adults form monogamous pairs and share a home range for most of the year. Juveniles disperse when they are 18–36 months old. They are usually relatively quiet and docile, emitting guttural snorts when disturbed. **Development** Males are sexually mature at 30–36 months, females at 24 months. The breeding season extends from February to May. A single young is born 15–17 days after conception. It attaches to one of 2 teats in the mother's forward-opening pouch, which it vacates 5–6 months later, riding on the mother's back until weaned at 8–9 months. Males have a lifespan of up to 12 years, and females to 17 years. **Diet** Leaves, fungi, lichen, bark and pine cones. **Habitat** Tall, wet sclerophyll forests and rainforests with a good understorey. **Traces** Claw marks on trees. Scats are about 10 mm across and 15–25 mm long, brown, comprising coarse plant fragments, often found on fallen logs and around trees. **Status** Low risk.

Weight **2.2–4.5 kg** Head–body **400–570 mm** Tail **340–420 mm**

FAMILY **PHALANGERIDAE** SPECIES *Trichosurus vulpecula*

COMMON BRUSHTAIL POSSUM

Often found living in roof spaces, the common brushtail possum is a rabbit-sized arboreal marsupial, usually silver-grey above and off-white to cream below. It has large pointed ears, dark patches on the muzzle and white patches below the ears. The tail is black, bushy, relatively short and moderately prehensile, with a naked area beneath the tip. A clawless opposing first toe on the hind foot helps it grip branches, while the second and third toes are fused with a long, split claw used for grooming. A north Queensland subspecies has short, coppery-coloured fur and Tasmanian animals are larger with denser fur. **Behaviour** Common brushtail possums are nocturnal and very agile climbers using their tail, sharp claws, and opposing toes to grasp branches. Tree hollows, logs, dense undergrowth, rabbit holes and roof spaces are used as dens. Adults are solitary for most of the year, and mark their den sites and home ranges with secretions from scent glands on the chin, chest and anal region. Individuals use several dens, moving between them during the year, and some may be used independently by other possums. Dens are shared only during the breeding season or by a mother and her young. Adults often travel long distances on the ground and establish home ranges as large as 7 ha for males and 5 ha for females. Dominant males have larger home ranges often overlapping those of several females. Young females establish a home range adjacent to their mother's, while young males disperse further afield. Their vocal repertoire is extensive and includes guttural coughs, sharp hisses, loud chattering and screeching. **Development** Females may give birth in their first year and males are fertile by the end of their second year. Breeding occurs year round in the north, with peaks in autumn and spring in other areas. In good conditions females may give birth twice in a year, producing a single young 16–18 days after mating. It attaches to one of 2 teats in the mother's forward-opening pouch, remaining attached for about 3 months and leaving the pouch at 4–5 months. Thereafter it is left in the den or rides on its mother's back until 7–9 months old. They may live to 13 years. **Diet** Leaves, shoots, fruits, flowers, grass and insects. **Habitat** Sclerophyll forests, rainforests, woodlands, mangroves and urban areas. **Traces** Claw marks on trees. Scats are red-brown to black, 3–10 mm wide and 20–30 mm long, deposited in groups or strings, sometimes containing insect remains. **Status** Low risk nationally, but endangered in the NT. Threatened by habitat fragmentation and fox predation.

WEIGHT **1.2–4.5 kg** HEAD–BODY **320–580 mm** TAIL **240–400 mm**

FAMILY **PHALANGERIDAE** SPECIES *Wyulda squamicaudata*

SCALY-TAILED POSSUM

A rabbit-sized arboreal marsupial and the only member of the genus, the scaly-tailed possum lives in rugged, rocky terrain and has affinities with the brushtail possums and the cuscus. It has pale grey fur on the back with a slightly darker stripe running from the crown to the rump. It is creamy-white below and sometimes rufous around the base of the prehensile tail, which is naked for most of its length with prominent scales. The head is quite long with a flattened pink nose, small ears and large bulging eyes. The hind feet have an opposing first toe for grasping branches, while the second and third toes are fused with a double claw used for grooming. **Behaviour** Scaly-tailed possums sleep by day in cool rock crevices or deep within rock piles. They emerge at night to feed on the ground and in trees where they are agile climbers, gripping branches with their hind feet and using the prehensile tail as a fifth limb to hold onto branches and support the body weight. If disturbed they descend immediately to the ground and shelter in rock fissures. They emit a bird-like chittering, chirping alarm call. They are solitary and both sexes live in overlapping home ranges averaging about 1 ha. They use multiple dens within the home range. **Development** They have a lifespan of up to 6 years in captivity. Females are sexually mature at about 2 years, males at 18 months. Females usually give birth in the dry season from March to August, although a mother with a young one in the pouch has been found in January. The single young remains in the mother's forward-opening pouch for 21–28 weeks, and is weaned at about 35 weeks. **Diet** Leaves, flowers, fruits, seeds and insects. **Habitat** Open and closed forests and vine thickets in rugged, rocky sandstone country with deeply-dissected boulders and more than 900 mm of rain annually. **Status** Low risk.

WEIGHT **1.2–2 kg** HEAD–BODY **290–400 mm** TAIL **255–350 mm**

FAMILY **POTOROIDAE** SPECIES *Aepyprymnus rufescens*

RUFOUS BETTONG

The largest bettong, distinguished by its hairy muzzle, the rufous bettong is reddish-brown to grey-brown flecked with light grey above, and lighter below with a very faint whitish hip stripe. Like the other bettongs it has a very long fourth toe on its hind foot and a double claw on its fused second and third toes for grooming. It has long curved claws on its forefeet for digging, and uses its semi-prehensile tail to carry nesting material. Females are larger than males. **Behaviour** Rufous bettongs forage on the forest floor at night, digging holes with their forefeet, and often feeding in small groups. When alarmed they stand upright, hold their arms stiffly by their sides, hiss, stamp their hind feet and hop quickly away, forelegs tucked against their sides. They sleep in spherical to cone-shaped nests made of grass collected in the mouth and carried in the curled-up tail. Clusters of nests are built often close to a feeding site. They have a single entrance and are constructed over shallow scrapes at the base of grass tussocks, under logs or rocks. Individuals may travel 4.5 km overnight and use up to 5 nests at any one time, abandoning them every month or so. Males have home ranges of 75–110 ha and females 45–60 ha. Their home ranges overlap, and some males associate exclusively with 1–2 females, guarding them from other male Potoroidae. Vocalisations include growls, grunts and low, warning hisses. **Development** Sexually mature at 11–12 months, they have a lifespan of 6–8 years and breed throughout the year. Females mate soon after giving birth, although the embryo remains dormant until the pouch is vacated. After a pregnancy of 22–24 days a single young is usually born, probably in the nest, and attaches to one of 4 teats in the mother's pouch for 7–8 weeks. The pouch is vacated at 14–16 weeks and the young suckles at foot for another 7 weeks. **Diet** Tubers, roots, herbs, fungi, insects, seeds, tree exudates and flowers. They can survive without drinking, except in droughts. **Habitat** Open forests and woodlands with a sparse or grassy understorey, to 1000 m. **Traces** Small conical diggings over large areas. Scats are oval, about 10 mm across and 10–20 mm long, containing coarse plant material. **Status** Low risk in Qld, but vulnerable in NSW. Threatened by land clearing, competition from rabbits and fox predation.

WEIGHT **3–3.5 kg** HEAD–BODY **370–390 mm** TAIL **335–400 mm**

Family **Potoroidae** | Species *Hypsiprymnodon moschatus*

Musky Rat-kangaroo

This rat-sized marsupial is the smallest macropod and the only one with an opposing, clawless first toe on the hind foot, enabling it to climb and grip branches. The fur is dense and soft, rich rufous-brown flecked with darker hairs above, paler below with patches of white to cream on the throat and chest. The tail is relatively short, dark brown with small non-overlapping scales. The head is long and slender with large rounded ears and a naked nose. The hind feet are relatively short with 5 toes, sharp curved claws and striated pads on the palms and soles to help them climb; the second and third toes are fused at the base. **Behaviour** Active during the day, they are generally solitary, foraging in the early morning and late afternoon on the forest floor and along creek banks. Groups of 2–3 are sometimes seen feeding on fallen fruit. Seeds are sometimes buried and consumed in the cooler months when food is scarce. They sleep at midday and at night in hollow logs, in rock piles or in nests of grass, ferns and lichen carried in the curled tail and placed in clumps of lawyer vines or between the buttresses of large trees, up to 1 m above ground. Individuals maintain small home ranges and use several nest sites while females build much larger maternity nests. They climb among fallen trees and branches and move with a slow hop on the ground with the tail held out behind, or run with a galloping gait, the hind feet moving forward outside the forelimbs. Courtship precedes mating by several days and includes standing erect, face-to-face, and touching the partner's head and neck with the forepaws. **Development** Sexually mature at 18–21 months, they have a lifespan of about 4 years. Males are sexually active from October to April, and most births occur between February and April. Usually 2 young are born and attach firmly to the 4 teats in the mother's pouch. They are evicted from the pouch in October and are then left in the nest. The mother suckles them for a few more weeks and they begin to accompany her on foraging expeditions, becoming fully independent by February. **Diet** Fruits, large seeds, some flowers, fungi, insects. Food is manipulated with the forepaws. **Habitat** Tropical rainforests near waterways, to 1500 m. **Traces** Scats are dark brown to black, roughly cylindrical, about 5 mm across and 5–20 mm long, containing fine fibrous particles and some insect remains. **Status** Low risk.

Weight **337–680 g** | Head–body **153–273 mm** | Tail **123–165 mm**

FAMILY **POTOROIDAE** SPECIES *Bettongia gaimardi*

TASMANIAN BETTONG

The Tasmanian bettong is now extinct on the mainland, but is still widespread in Tasmania. The size of a small rabbit, it has a semi-prehensile tail used to carry nesting material, well-developed upper canine teeth, and upper and lower incisor teeth that bite against each other. The broad head has rounded ears and a naked nose. The fur is coarse with soft underfur, brownish-grey flecked with dark brown to black above, and greyish-white below. The tail is well-furred, dark brown towards the end with a white tip. The short muscular forelimbs have forward-pointing toes with long, curved claws for digging. The hind feet have no first digit, the second and third toes are fused with a double claw used for grooming, while the fourth toe is much longer than the others. **Behaviour** Nocturnal and usually solitary, Tasmanian bettongs have large home ranges of 35–55 ha, and may travel up to 1.5 km from the nest to a feeding site, where they sniff out and dig up underground fungi. They sleep by day in ovoid nests about 300 mm long and 200 mm wide with a small entrance, built from densely-woven dry grass and bark strips carried in the curled tail and manipulated with the snout. Nests are constructed over shallow depressions dug beneath shrubs, grass tussocks or logs. Up to 6 nests may be built in the home range; one may be used for a month or more and then abandoned for a similar period before being reoccupied. They have anal scent-producing glands, probably used to mark the nest site. **Development** Sexually mature at 8–12 months, Tasmanian bettongs have a lifespan of 3–5 years in the wild. They breed throughout the year and females mate soon after giving birth, although the embryo remains dormant until the pouch is vacated. After a pregnancy of 21–22 days, a single young is born and attaches to one of 4 teats in the mother's pouch. The pouch is vacated at 14–16 weeks and the young suckles at foot for a further 7–8 weeks. **Diet** Mostly underground fungi, with some seeds, fruits, roots, bulbs and insects. **Habitat** Dry sclerophyll forests and woodlands with an open, grassy understorey, to more than 1000 m, with an annual rainfall of 500–750 mm. **Traces** Small conical diggings and oval scats about 10 mm across and 10–20 mm long, containing fine soil particles. **Status** Low risk.

WEIGHT **1.2–2.3 kg** HEAD–BODY **310–350 mm** TAIL **285–350 mm**

FAMILY **POTOROIDAE** SPECIES *Bettongia lesueur*

BURROWING BETTONG OR BOODIE

This rabbit-sized, kangaroo-like marsupial was, before European colonisation, widespread across western, southern and central Australia. Today it exists only on Bernier and Dorre Islands in Shark Bay, Boodie and Barrow Isand off the Pilbara coast of WA, and in fenced sanctuaries on the mainland. Small and thickset, it has powerful hind legs, stumpy forepaws and short, dense, soft fur, grey to yellow-grey above and light grey below. The pointed head has short rounded ears. The tail is fat, sparsely-haired and weakly prehensile, and sometimes has a white tip. **Behaviour** Burrowing bettongs emerge after sunset and forage independently until sunrise. They move slowly using their hind legs, sniffing the ground, digging up roots and tubers, and often climbing into low shrubs for flowers or fruit. They shelter in simple burrows with a short, shallow, curving tunnel and 1–2 entrances; or share large communal warren systems with dozens of entrances and deep, interconnecting tunnels. Nests made from grass, probably carried in the curled tail, are constructed in the burrows, and up to 50 individuals may share a warren, forming social groups comprising a male and one or more females. Adult males never share burrows with other males and vigorously defend their harems against rivals. Females share burrows with their offspring and sometimes with other females. They are very vocal, uttering a variety of squeals, hisses and grunts. **Development** Burrowing bettongs reach sexual maturity in their first 12 months and have a lifespan of at least 3 years. They breed throughout the year, producing up to 3 young per year. A single young (rarely twins) is born 21 days after mating. The tiny, hairless newborn attaches to one of the 4 teats in the mother's pouch, and remains in the pouch for about 4 months. It is weaned 3–10 weeks later. Females mate the day after giving birth although the embryo remains dormant until the pouch is vacated. **Diet** Seeds, fruit, flowers, tubers, roots, fungi, succulent leaves and grasses, termites. **Habitat** Most arid and semi-arid habitats, including coastal dunes and hummock grasslands on the islands. Previously found in various habitats on the mainland including open forests, scrubs and spinifex grasslands. **Traces** Small conical diggings; spoil heaps beside warrens. **Status** Vulnerable. Threatened by land degradation and predation by feral animals.

WEIGHT **0.4–1.8 kg** HEAD–BODY **280–400 mm** TAIL **215–300 mm**

FAMILY **POTOROIDAE** SPECIES *Bettongia penicillata*

BRUSH-TAILED BETTONG OR WOYLIE

This squat, rabbit-sized, kangaroo-like marsupial has a prehensile tail used to gather nesting material, well-developed upper canine teeth and upper and lower incisor teeth that bite against each other. It is sandy-brown to yellow-grey flecked with white above, and pale grey-brown to white below, with a black bushy tail tip. It has a broad head with a flattened, naked nose and pointed ears. The short, muscular forelimbs have forward-pointing toes with spatulate claws used for digging. The hind feet have no first digit, the second and third are fused with a double claw used for grooming, and the fourth is much longer than the others.

Behaviour Brush-tailed bettongs forage at night, digging in the soil and leaf litter for food. They hop away quickly if disturbed, with the head low, the back arched and the tail held out straight behind. They sleep in well-concealed domed nests about 200 mm across, made from shredded bark lined with grass carried in the curled tail. The nests are built over shallow depressions dug beneath shrubs or other cover. They are solitary except during the breeding season, and utilise large home ranges of 20–40 ha, incorporating several exclusive nests and a feeding area sometimes shared with neighbouring bettongs. **Development** Female brush-tailed bettongs are sexually mature at 5–6 months, males at 9–12 months, and they have a life expectancy of 4–6 years. They breed year round, producing 2–3 litters per year. After a pregnancy of 21 days the newborn attaches to one of 4 teats in the mother's pouch, which it vacates at about 14 weeks, suckling at foot for another 60–70 days and sharing the mother's nest until the next young leaves the pouch. Mating occurs soon after birth, although the embryo remains dormant until the pouch is empty. **Diet** Mostly underground fungi, supplemented with bulbs, tubers, seeds, gum exudate and insects. They can survive without drinking. **Habitat** Dry sclerophyll forests and woodlands with good ground cover, to about 300 m. **Traces** Small, cone-shaped diggings. Scats are oval to cylindrical, about 8 mm across and 10–20 mm long, containing fine soil particles and fungal spores. **Status** Low risk but conservation dependent.

WEIGHT **1.1–1.6 kg** HEAD–BODY **300–380 mm** TAIL **290–360 mm**

FAMILY **POTOROIDAE** SPECIES *Potorous longipes*

LONG-FOOTED POTOROO

About the size of a rabbit, this shy rat-kangaroo is one of the rarest mammals in Australia. It is has a dense coat of soft grey-brown fur, becoming paler on the belly and feet. It is larger than the long-nosed potoroo with longer hind feet bearing long toes with strong claws and a leathery pad on the sole. It has a long nose, an almost hairless tail, short muscular forelimbs with strong claws for digging and a double claw on the fused second and third toes of the hind feet used for grooming. Males are larger than females. **Behaviour** Long-footed potoroos are generally solitary, although males and females may forage in pairs. They are active at night and shelter during the day in a simple nest scraped in the ground beneath dense clumps of ferns or wiregrass. They forage in open areas at night, digging small conical pits with the forefeet, looking for fungi. They live in home ranges varying from 10–40 ha. Male home ranges overlap with a number of female home ranges, but not with those of other males. Vocalisations are restricted to low kissing sounds uttered between mothers and their young or when under stress. **Development** Long-footed potoroos reach sexual maturity at 2 years and have a lifespan of more than14 years in captivity. They breed throughout the year with peaks in winter, spring and early summer, and may produce 2–3 young per year. A single young is born about 38 days after mating and attaches to one of 4 teats in the mother's well-developed pouch. Young vacate the pouch at 4 months and live in the mother's home range until about 12 months old. **Diet** Fungi supplemented with other plant matter and some invertebrates. **Habitat** Moist sclerophyll forests with a dense understorey from 150–1000 m, also warm temperate rainforests. **Traces** Similar to the long-nosed potoroo (page 92). **Status** Endangered. Threatened by habitat loss and predation by feral animals.

WEIGHT **1.6–2.2 kg** HEAD–BODY **280–415 mm** TAIL **315–325 mm**

FAMILY **POTOROIDAE** SPECIES *Potorous tridactylus*

LONG-NOSED POTOROO

This squat, rabbit-sized, kangaroo-like marsupial has a prehensile tail used to gather nesting material, well-developed upper canine teeth and upper and lower incisor teeth that bite against each other. The fur is grey to brown above (dark rufous-brown in Tasmania) and paler below. It has a long, tapering nose with a naked tip, rounded ears and a scaly tail with a furry base. The forelimbs are short and muscular and have forward-pointing toes with spatulate claws used to dig for food. The second and third toes of the hind feet are short and fused with a double claw used for grooming, the fourth toe is long and the fifth short. Males are larger than females. **Behaviour** Long-nosed potoroos are generally nocturnal, although in winter they may forage by day. They search for food on the forest floor, moving slowly with a short hop, digging up underground fungi and roots with their clawed forefeet. They are solitary outside the breeding season and tend to stay in the same area, living in a home range of around 5 ha. They rarely venture far from cover and sometimes gather in small groups. If disturbed they leap quickly away on their hind legs and use their forelimbs to change direction. They sleep by day in simple nests of grass and other vegetation carried in the curled tail and placed in scrapes below dense scrub, grass tussocks or grass trees. In bush fires they seek refuge in the burrows of other animals. **Development** Sexually mature at 12 months, long-nosed potoroos have a lifespan of around 4–5 years in the wild (7 years in captivity), and breed year-round with peaks in early spring and summer. A single young is born 38 days after mating and attaches firmly to one of 4 teats in the mother's pouch, leaving the pouch at about 4 months and suckling at foot for 5–6 weeks. Mating may occur soon after birth, but the embryo remains dormant until the pouch is empty. **Diet** Mostly fungi, supplemented by insects, roots, seeds and fruit. **Habitat** Coastal heaths, rainforests, sclerophyll forests and woodlands with thick ground cover and occasional open areas, to 650 m. **Traces** Cone-shaped diggings up to 250 mm deep. Scats are deposited near feeding sites. They are oval to cylindrical, 5–10 mm across, comprising fine particles of soil, spores and plant material. **Status** Vulnerable. Threatened by land clearing and predation by feral animals.

WEIGHT **660–2070 g** HEAD–BODY **300–400 mm** TAIL **150–265 mm**

FAMILY **MACROPODIDAE** SPECIES *Dendrolagus bennettianus*

BENNETT'S TREE-KANGAROO

Similar in many respects to Lumholtz's tree-kangaroo, this arboreal marsupial of the rainforests of northeastern Australia is dark brown above and light fawn below with a rusty-brown area on the back of the head and shoulders. The forehead and muzzle are greyish, the feet, hands and base of the tail are black, the long tail has a bushy tip and a light brown patch on the upper surface. The ears are small and rounded. It has stout, muscular forelimbs and relatively short hind feet with toes all about the same length and granular soles. The claws are strong and curved. The second and third toes of the hind feet are fused with a double claw used for grooming. Males are much larger than females.

Behaviour Predominantly nocturnal, Bennett's tree-kangaroos spend most of the day sitting high up in the canopy, often basking in the sun in winter, hidden from view by vines and vegetation, and sleeping crouched on a branch. Wary and cryptic, they climb through the canopy at night, gripping branches with their clawed forefeet and balancing with the hind feet. They can leap 9 m between trees and back down tree trunks to the ground where they walk or run quadrupedally or hop rapidly, often travelling long distances. Adults defend a discrete home range centred on large trees used for resting, and males will fight ferociously to repel intruders. Male home ranges may be up to 25 ha and often overlap those of 2–3 females. Family groups of male, female and young at foot often feed in close proximity at night. They growl when alarmed and females call their young with soft trumpeting sounds.
Development They have a lifespan of 20 years or more in captivity and females become sexually mature at about 2 years. They breed year-round and females give birth to a single young which attaches to one of the 4 teats in her pouch. The young leave the pouch at 8 months and stay with the mother until they are up to 2 years old. **Diet** Leaves supplemented by fruit. **Habitat** Highland and lowland tropical rainforests and vine forests.
Status Rare. Threatened by logging and land clearing.

WEIGHT **8–13.7 kg** HEAD–BODY **690–750 mm** TAIL **730–840 mm**

FAMILY **MACROPODIDAE** SPECIES *Dendrolagus lumholtzi*

LUMHOLTZ'S TREE-KANGAROO

Found in the rainforests of northeastern Australia, this unusual kangaroo varies in colour from pale creamy-brown to grey or rusty-brown with lighter flecks on the rump. It has black paws and a pale band across the forehead and down the sides of the face. The ears are very small and rounded and the tail is long and thick but not prehensile. It has shorter limbs than the ground-dwelling kangaroos with stout, muscular forelimbs armed with strong, sharp, curved claws to help it climb. The toes on the hind feet are all about the same length with uniformly granular soles for extra grip and strong, curved claws. A double claw on the second and third fused toes is used for grooming. Males are larger than females.

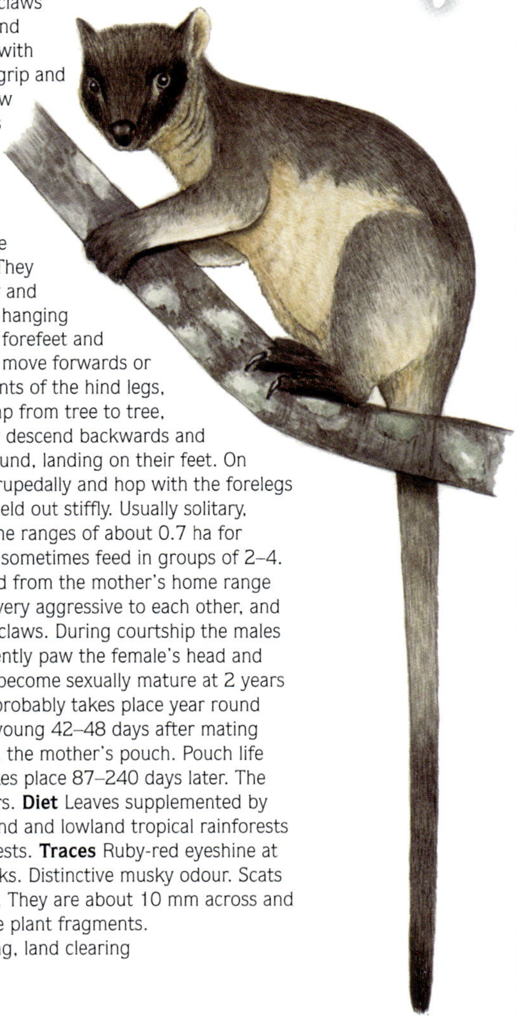

Behaviour Lumholtz's tree-kangaroos are nocturnal, sleeping by day crouched on a branch in the crown of a tree, often in the sun. They feed in the mid-level of the canopy and understorey, climbing with the tail hanging loosely, gripping with their clawed forefeet and balancing with the hind feet. They move forwards or backwards with alternate movements of the hind legs, hop along broad branches and jump from tree to tree, crashing through the canopy. They descend backwards and can jump 10 m or more to the ground, landing on their feet. On the ground they walk or run quadrupedally and hop with the forelegs tucked into the body and the tail held out stiffly. Usually solitary, they occupy small overlapping home ranges of about 0.7 ha for females and 1.8 ha for males, and sometimes feed in groups of 2–4. Juvenile males are probably evicted from the mother's home range and forced to disperse. Males are very aggressive to each other, and can inflict fatal wounds with their claws. During courtship the males make a soft clucking sound and gently paw the female's head and shoulders. **Development** Females become sexually mature at 2 years and males at 4.6 years. Breeding probably takes place year round and females give birth to a single young 42–48 days after mating which attaches to one of 4 teats in the mother's pouch. Pouch life is 246–275 days, and weaning takes place 87–240 days later. The young stay at heel for up to 2 years. **Diet** Leaves supplemented by fruits and flowers. **Habitat** Highland and lowland tropical rainforests and associated wet sclerophyll forests. **Traces** Ruby-red eyeshine at night. Deep scratches on tree trunks. Distinctive musky odour. Scats are sometimes deposited on rocks. They are about 10 mm across and 10–25 mm long, containing coarse plant fragments.
Status Rare. Threatened by logging, land clearing and feral dogs.

WEIGHT **3.7–10 kg** HEAD–BODY **500–600 mm** TAIL **600–736 mm**

Family **Macropodidae** Species *Lagorchestes conspicillatus*

Spectacled Hare-wallaby

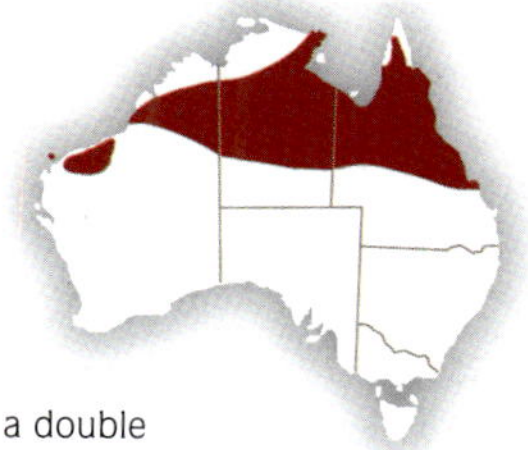

The spectacled hare-wallaby is a stout, short-necked marsupial of the tropical grasslands. It has a superficial resemblance to the European hare, and has a rufous-brown ring surrounding each eye. It is brown flecked with white above, and white below, with white stripes on the hips. The tail is sparsely-haired and darker towards the tip. The face is square, viewed from the front, with a black nose and pointed ears. The long hind feet have no first toe, the second and third are fused with a double claw used for grooming, and the fourth toe is much longer than the others. **Behaviour** Spectacled hare-wallabies are nocturnal, emerging at night to graze, and are generally solitary although several animals may be seen feeding together. They shelter from the daytime heat in cool hides tunnelled into grass tussocks, spinifex hummocks, or in clumps of porcupine grass. In Acacia woodlands they shelter under shrubs beneath shady trees. Several hides are used in a home range of 8–10 ha. They are powerful jumpers and move with a fast, hopping gait. Vocalisations include a warning hiss and a soft clicking sound made by males around females on heat. They conserve water in hot dry conditions by raising their body temperature and eating less food, hence reducing the amount of water needed for cooling and digestion. **Development** Sexually mature at 12 months, spectacled hare-wallabies breed year-round, with peaks in March and September on Barrow Island. A single young is born 29–31 days after mating and attaches firmly to one of the 4 teats in the mother's pouch. The pouch is vacated at about 5 months. Mating takes place soon after giving birth, although development of the embryo is delayed until the pouch is empty, or in drought conditions until food is more plentiful. **Diet** New growth on shrubs, grasses and herbs; spinifex leaves. They do not need to drink. **Habitat** Hummock grasslands, open forests, open woodlands and tall shrublands. **Traces** Scats are slightly flattened, dark brown, oval to oblong pellets; about 10 mm across and 10–20 mm long; containing fairly coarse plant fragments; usually found in groups of about 4–8. **Status** Vulnerable. Threatened by predation by feral animals and habitat degradation.

Weight **1.6–4.5 kg** Head–body **400–470 mm** Tail **370–390 mm**

FAMILY **MACROPODIDAE** SPECIES *Lagorchestes hirsutus*

RUFOUS HARE-WALLABY OR MALA

Before European settlement, this small rabbit-sized wallaby was one of the most abundant and widespread macropods of central Australia. It is now confined to Bernier and Dorre Islands in Shark Bay, WA, and protected colonies on the mainland. It has a shaggy appearance with long, soft, rufous-brown fur on the hindquarters, chest and abdomen, grading to pale fawn or white on the belly, with a greyish-white tint on the head and back. The paws, feet and tail are darker. It has large beady black eyes, large pointed ears and a naked nose. **Behaviour** Solitary and nocturnal they shelter during the day in small scrapes hidden beneath bushes or spinifex clumps. In the intense summer heat they dig burrows about 1 m long and 30 cm deep with a single entrance. They emerge after sunset and conduct most of their activities early in the night, avoiding other hare-wallabies, although there is evidence of social hierarchies in males and females. They maintain large, sparsely-used home ranges of around 60 ha in dense spinifex habitat, adjoining small, well-used feeding sites on more open saline flats and areas regenerating after fire. Male home ranges overlap those of a number of females, and males have been observed chasing ovulating females. Males spray urine to mark their home ranges. If alarmed they make a loud squeak then hiss as they bolt for cover, hopping on their powerful hind legs. When grazing they move slowly on all fours. **Development** Females are sexually mature at 5–18 months and males at 14–20 months, and have a lifespan of 8 years or more. They are able to breed year round with birth peaks after rain. A single young is born and suckles in the pouch for about 4 months. **Diet** Grasses, seeds, fruits, herbs, sedges, succulent shrubs and insects. **Habitat** Low shrublands, heaths and spinifex hummock grasslands. **Traces** Scats are 10–15 mm long and 5–10 mm wide, pointed at one end, light to dark brown, containing fairly coarse plant matter. They are found in groups of 4–8 around feeding and shelter sites and on paths through the undergrowth. **Status** Vulnerable. Threatened by predation by feral animals and habitat degradation.

WEIGHT **0.7–2 kg** HEAD–BODY **310–190 mm** TAIL **240–305 mm**

FAMILY **MACROPODIDAE** SPECIES *Macropus agilis*

AGILE WALLABY

The agile wallaby is the most common macropod found along the coast of tropical Australia. Sandy brown above and light buff to whitish below, it has a whitish stripe along the cheek and thigh, a dark brown stripe down the forehead, a long black-tipped tail and black-edged ears. The long pointed face has a partially naked muzzle. Like other macropods the hind feet lack a first digit, the second and third are fused with a double claw for grooming, while the fourth is much longer than the others. Males are much larger than females.
Behaviour Active during the late afternoon and night, agile wallabies rest in the heat of the day in dense vegetation. They are gregarious animals, living in groups of up to 10 individuals and congregating in much larger mobs when food is short. During the breeding season dominant males monopolise females on heat, staying close to them and keeping other males away. Agile wallabies are very alert and nervous. When alarmed they thump their hind feet and hop rapidly away with the head held high, the tail horizontal and the forearms extended.
Development Agile wallabies have a lifespan of up to 14 years. Females are sexually mature at 12 months, males at about 14 months, and breeding takes place throughout the year with a peak between May and August. Females mate soon after giving birth, producing a single embryo that remains dormant until the pouch is empty. After a pregnancy of 29–31 days, the newborn attaches firmly to one of 4 teats in the mother's pouch. The infant stays in the pouch for 7–8 months and then suckles at foot until weaned at 10–12 months. **Diet** Grass, sedges, leaves, fruits and roots dug up with the forepaws.
Habitat Along creeks in open forests and adjacent grasslands, and coastal sand dunes. **Traces** Scats are oval, square or round pellets; 15–25 mm across; containing coarse plant fragments. **Status** Low risk.

WEIGHT **9–29 kg** HEAD–BODY **593–850 mm** TAIL **587–840 mm**

FAMILY **MACROPODIDAE** SPECIES *Macropus antilopinus*

ANTILOPINE WALLAROO

A large marsupial of the tropical northern woodlands, the antilopine wallaroo resembles the grey kangaroos in appearance and behaviour, and derives its name from its long, fine, supposedly antelope-like fur. It is reddish sandy-brown above and very pale brown to white below, often with pale patches on the legs and tail. Females may be bluish-grey. The paws are black, the nose is naked and the tail relatively short and thick. The hind feet have no first digit, the second and third are fused with a double claw for grooming, and the fourth is much longer than the others. Males are much larger than females. **Behaviour** Antilopine wallaroos are generally active at night, although on overcast or wet days they sometimes forage during daylight hours. They rest in the shade of trees, shrubs or rocks, usually near a waterhole, moving out to graze in the late afternoon. They are gregarious and outside the breeding season they gather in single sex groups comprising 3–5 large males and 5–10 females. Larger groups of up to 30 animals gather for safety if they have been disturbed by humans or dingos. Group membership changes frequently. Males groom each other regularly and establish dominance hierarchies. When alarmed they hiss and thump the ground with their hind feet as they hop rapidly away. In the breeding season males are often seen alone, probably searching for female mates, and have been recorded travelling within a home range as large as 76 ha. **Development** Antilopine wallaroos have a lifespan of up to 16 years and are thought to breed throughout the year, with a peak of births in the early dry season and towards the end of the wet season. A single young is born some 34 days after mating and attaches firmly to one of the 4 teats in the mother's pouch, vacating the pouch permanently at about 38 weeks, thereafter suckling at foot until about 12 months old. **Diet** Grass and leaves. **Habitat** Open savanna woodlands and open monsoonal forests, usually in flat or undulating country. **Traces** Scats are oval, square or round pellets; 15–25 mm across; containing coarse plant fragments. **Status** Low risk.

WEIGHT **16–60 kg** HEAD–BODY **775–1200 mm** TAIL **675–890 mm**

Family **Macropodidae** Species *Macropus bernardus*

Black Wallaroo

This is the smallest and stockiest of the wallaroos. Males are sooty-brown to black while females are pale grey to grey-brown. The paws, feet and base of the tail are black in males and dark brown in females. The muzzle is square with a black, completely naked nose. The hind feet have no first digit, the second and third are fused with a double claw for grooming, and the fourth is much longer than the others. Males are larger than females. **Behaviour** The black wallaroo is shy and generally solitary, although an adult male and female with a large young are sometimes seen together. Males have also been observed fighting other males over females. They shelter during the day in caves and sometimes in the shade of rocks or trees and in hummock grass, emerging to feed at dusk or on overcast days. They use a range of shelter sites in a home range. If disturbed they jump up and hop rapidly away leaping over rocks and cliffs until out of sight. They occasionally come down from the escarpment at night to drink at springs or waterholes and to graze, often with euros or antilopine wallaroos. **Development** Black wallaroos have a lifespan of 11 years or more and are able to breed throughout the year in good conditions. Females give birth to a single young about 34 days after mating. The newborn attaches firmly to one of the 4 teats in the mother's pouch, remaining attached for another 4 months. The young leaves the pouch at about 6 months and then suckles at foot. **Diet** Grass, leaves, possibly fruit, seeds, flowers and roots. **Habitat** Closed forests to open forests, heaths and hummock grasslands in areas characterised by large boulders on rugged sandstone escarpments. **Status** Low risk.

Weight **13–22 kg** Head–body **595–725 mm** Tail **545–840 mm**

FAMILY **MACROPODIDAE** SPECIES *Macropus dorsalis*

BLACK-STRIPED WALLABY

Very shy and seldom venturing far from cover, the black-striped wallaby shares part of its habitat with the red-necked wallaby, with which it is often confused. It can be distinguished by the black stripe running down the centre of its back, from neck to rump and its white thigh stripes. It has sandy to reddish-brown fur, paler on the flanks and grading to almost white below, and has white cheek patches. The tail is sparsely-haired and scaly. The hind feet have no first digit, the second and third are fused with a double claw for grooming, and the fourth is much longer than the others. Males are much larger than females. **Behaviour** Active mainly at night, black-striped wallabies feed from dusk to dawn in grassy areas with some tree cover, seldom venturing far from cover. They flee if disturbed, moving with a distinctive short hopping gait with the back hunched, the head low and forearms extended forward and out from the body. During the day they rest in dense vegetation in groups of 20 or more of mixed age and sex. They travel in single file from their resting site along established runways to feeding sites in more open areas where they feed in close proximity. They stay within a large home range of around 91 ha. Adults groom and lick each other frequently and are often seen sparring with each other, probably to establish and maintain dominance hierarchies. Aged males live a solitary existence. **Development** Black-striped wallabies have a life span of up to 20 years. Females are sexually mature at 14 months, males at 20 months. They breed year round and females mate soon after giving birth, although the embryo remains dormant until the pouch is empty. A single young is born after a pregnancy of 33–35 days. The newborn attaches firmly to one of the 4 teats in the mother's pouch, which it vacates at 30 weeks, thereafter suckling at foot until weaned at 10–12 months. **Diet** Native grasses, herbs and shrubby vegetation. **Habitat** Closed and open forests and woodlands with a dense understorey, including rainforest margins, brigalow scrub and lantana thickets. **Traces** Scats are oval to round or oblong pellets, often pointed at one end, about 10–15 mm across and 15–20 mm long, containing coarse plant fragments. **Status** Low risk in Qld, but endangered in NSW. Threatened by land clearing, grazing and predation by feral animals.

WEIGHT **6–20 kg** HEAD–BODY **530–820 mm** TAIL **540–830 mm**

FAMILY **MACROPODIDAE** SPECIES *Macropus eugenii*

TAMMAR WALLABY

One of the smallest wallabies, the tammar wallaby has long soft fur, dark grey-brown flecked with light grey above, with reddish-brown patches on the flanks and limbs. It is paler below with a white muzzle, a white stripe on the cheek and a dark stripe down the forehead. The ears are slightly pointed and the nose is naked. The hind feet have no first digit, the second and third are fused with a double claw, and the fourth is much longer than the others. Males are larger with much larger forelimbs and wider claws than females. **Behaviour** Tammar wallabies are usually active for a few hours after dusk and again for a few hours before dawn. They sleep by day in dense, low vegetation, and move off after dusk, following established runways to grassy feeding sites up to 1.2 km away, and returning before dawn. They are generally solitary with defined overlapping home ranges of up to 42 ha, although they sometimes feed with 1–2 others. Males are aggressive during the breeding season when they congregate around females on heat and try to dislodge a successful copulating male. A dominant male will guard a female from other males for up to 8 hours after mating. **Development** Tammar wallabies have a lifespan of up to 14 years, although juveniles have a high mortality rate in their first summer. Females become sexually mature while still suckling at about 9 months, while males mature at about 24 months. Young are born in late January and early February, and females mate 1–2 hours after giving birth. The embryo typically remains dormant until late December and a single young is born 25–28 days later. The newborn attaches firmly to one of 4 teats in the mother's pouch. It leaves the pouch permanently at 8–9 months, and suckles at foot until 10–11 months. **Diet** Native grasses, herbs, shrubs and seedlings. In droughts they can survive by drinking seawater. **Habitat** Dry sclerophyll forests and woodlands with a dense understorey, mallee thickets, coastal scrubs and heathlands. **Traces** Scats are round, oval or oblong pellets; about 10–15 mm across and 10–25 mm long; containing coarse plant fragments. **Status** Low risk nationally, but endangered in mainland SA. Threatened by land clearing and predation by feral animals.

WEIGHT **5–10 kg** HEAD–BODY **520–680 mm** TAIL **330–450 mm**

FAMILY **MACROPODIDAE** SPECIES *Macropus fuliginosus*

WESTERN GREY KANGAROO

Similar in many respects to the eastern grey kangaroo, this species has light to dark chocolate-brown fur, often flecked with grey above, and paler below. It has dark brown to black paws, feet and tail tip, and buff patches on the legs and forearms. It has a finely-haired muzzle and large ears fringed with white hairs. The hind feet have no first digit, the second and third are fused with a double claw for grooming, and the fourth is much longer than the others. Males are much larger than females. **Behaviour** Western grey kangaroos rest by day in the shelter of trees and shrubs, and feed from late afternoon to early morning. Mobs of 40–50 range within a home range of up to 8 sq km. Males have a strong, distinctive odour and establish dominance hierarchies by fighting and ritualised acts such as tree-gouging with the front claws. They range widely searching for females on heat. Females usually congregate with female relatives, form their own dominance hierarchies and will only mate with dominant males. Old males are usually solitary. When alarmed they make guttural coughs and thump their hind feet, hopping away with the body erect and the tail curved up. **Development** They have a lifespan of up to 20 years in the wild. Females are sexually mature at about 16 months, males at 20 months. They breed year round with a peak of births from September to March. Females mate 2–10 days after the pouch is vacated. After a pregnancy of 30 days the single newborn (rarely twins) attaches firmly to one of 4 teats in the mother's pouch, which it vacates at about 46 weeks, suckling at foot until about 17 months old. **Diet** Native grasses, herbs, leaves, tree bark and shrubs. **Habitat** Dry open forests, woodlands, open scrubs, wet and dry heaths, grasslands. **Traces** Similar to the eastern grey kangaroo (page 103. **Status** Low risk.

WEIGHT **15–66 kg** HEAD–BODY **520–1225 mm** TAIL **425–1000 mm**

FAMILY **MACROPODIDAE** SPECIES *Macropus giganteus*

EASTERN GREY KANGAROO

The eastern grey kangaroo is a large, robust marsupial with soft, deep fur varying from light silver-grey to dark grey flecked with light grey above, and usually paler below. Females have a white chest. The forehead is grey, while the paws, feet and tail tip are dark grey to black. It has a hairy nose and no first digits on the hind feet; the second and third toes are fused with a double claw for grooming, and the fourth is much longer than the others. Males are larger than females. **Behaviour** Predominantly nocturnal, eastern grey kangaroos rest in the shade and feed from late afternoon to early morning, often in mixed sex mobs of 10 or more, each mob occupying a home range of up to 5 sq km. Males range widely searching for females on heat and establish dominance hierarchies by fighting and ritualised acts including grass-pulling with the forefeet. Females usually congregate with female relatives and will only mate with dominant males. Old males are usually solitary. Clucking sounds are made between mother and young and during courtship. When alarmed they make guttural coughs and thump their hind feet as they hop away with the body erect and the tail curved up. They can also swim to avoid predators. **Development** They have a lifespan of up to 25 years. Females reach sexual maturity at about 22 months and males at about 43 months. Breeding occurs year-round with a peak of births in spring and early summer or when conditions improve after drought. Females mate about 11 days after the pouch is vacated (in good seasons they mate when the pouch young is more than 4 months old) and the embryo remains dormant until the pouch is empty. After a pregnancy of 33–38 days the newborn attaches firmly to one of 4 teats in the mother's pouch, which it vacates at about 11 months, suckling at foot until 18 months. Twins have been recorded. **Diet** Grasses, herbs, leaves and low shrubby vegetation. **Habitat** Dry sclerophyll forests, woodlands and low open scrub with grassy areas, from sea level to more than 1000 m. **Traces** Water digs in dry desert watercourses. Scats are oval to squarish pellets, about 10–20 mm across and 20–30 mm long, containing fine-textured green to brown plant fragments, and sometimes deposited in large clumps. **Status** Low risk.

WEIGHT **20–85 kg** HEAD–BODY **500–1212 mm** TAIL **430–1200 mm**

FAMILY **MACROPODIDAE** SPECIES *Macropus irma*

WESTERN BRUSH WALLABY

The western brush wallaby is a grazing kangaroo of the open forests and woodlands of southwestern Australia. It has a pale grey back sometimes tinged with brown, distinctive black and white ear margins and a white cheek stripe. The long tail has a crest of black hair, and some animals have faint pale bars across the back and tail. The hind feet have no first digit, the second and third are fused with a double claw for grooming, and the fourth is much longer than the others. **Behaviour** Western brush wallabies have not been studied in detail. Unlike many other kangaroos they are most active in the early morning and late afternoon, when they feed in open grassy areas. The remainder of the day is spent resting in the shade or sleeping among low trees or in shrubby thickets, alone or in pairs. Western brush wallabies are well-adapted to the open country and are able to move very fast and change direction suddenly, hopping with the head held low and the tail straight out behind. **Development** Females appear to give birth to a single young in April or May. The newborn attaches firmly to one of the 4 teats in the mother's pouch. Young leave the pouch permanently at about 6–7 months and then suckle at foot. **Diet** Native grasses, leaves of succulents, cycads and a variety of other shrubs. They can survive without drinking. **Habitat** Open forests and woodlands with grassy areas and shrubby thickets; mallee and heaths. **Status** Low risk.

WE GHT **7–9 kg** HEAD–BODY **830–1530 mm** TAIL **540–970 mm**

FAMILY **MACROPODIDAE** SPECIES *Macropus parma*

PARMA WALLABY

A cryptic inhabitant of forests with thick shrubby understoreys, the parma wallaby was thought to be extinct in Australia until it was rediscovered near Gosford in 1967. It has thick fur, grey-brown to dark brown above with a dark stripe from the head to the mid-back, and white stripes along the upper lip, throat, chest and belly. Many animals have a white-tipped tail. The hind feet have no first digit, the second and third are fused with a double claw, the fourth is much longer than the others. Males are larger than females. **Behaviour** Active mainly at night, parma wallabies are solitary animals. They rest under shrubs in dense vegetation by day and emerge at or just before dusk to feed, travelling along established runways to small grassy areas where they feed alone or sometimes with 1–2 others. They hop close to the ground, almost horizontally, with their forearms tucked into their sides and the tail curved up slightly. Courtship behaviour includes the male pawing the female's rump and holding her head on his chest with his forepaws. Vocalisations include clucking and warning hisses. **Development** Females are sexually mature at 12 months, males at 20–24 months. The breeding season extends throughout the year with most births from February to June. Parma wallabies usually mate 45–105 days after giving birth, although the embryo remains dormant until the pouch is empty. After a pregnancy of 33–36 days the newborn attaches firmly to one of 4 teats in the mother's pouch. It makes its first excursions from the pouch at 23–25 weeks and vacates the pouch at about 30 weeks, thereafter suckling at foot until about 10 months old. **Diet** Grasses and herbs. **Habitat** Rainforests and sclerophyll forests with a dense understorey and grassy sites, usually in montane areas. **Traces** Scats are round, oval or oblong pellets; about 10–15 mm across and 10–20 mm long; containing coarse plant fragments. **Status** Vulnerable. Threatened by fox predation and logging.

WEIGHT **3.2–5.9 kg** HEAD–BODY **445–530 mm** TAIL **405–545 mm**

FAMILY **MACROPODIDAE** SPECIES *Macropus parryi*

WHIPTAIL WALLABY

A slim, graceful kangaroo of the open forests, the whiptail wallaby is known for its long, slender tail. It has a triangular face with large brown and white ears, a dark brown forehead and a contrasting white cheek stripe. It is white below with a light brown stripe along the neck and shoulder, a white stripe on the hip, dark paws and a dark tip to the tail. The winter coat is light grey above and the summer coat brownish-grey. The hind feet have no first digit, the second and third are fused with a double claw for grooming, and the fourth is much longer than the others. Males are much larger than females. **Behaviour** Whiptail wallabies feed mainly around dawn and early morning, resting in the shade until late afternoon and feeding into the night. In winter they feed at any time of day. They are very gregarious and live in mobs of 30–50 animals with a large stable home range of up to 100 ha or more, abutting those of neighbouring mobs. Individuals or small groups have overlapping home ranges within this area and dominance hierarchies are established by aggression and rituals such as grass-pulling with the forepaws. A female on heat is followed by a group of males, but the dominant male keeps the subordinates at a distance. When alarmed they scatter erratically, thumping the ground with the hind feet as they hop away, taking a zig-zag course with the back and tail almost horizontal. Vocalisations include a soft submissive cough, soft clucking during courtship and a threatening growling hiss. **Development** Females are sexually active at 18–24 months, males at 2–3 years and have a lifespan of up to 10 years. Births occur throughout the year with a peak from March to July. Females mate near the end of the young's pouch life, although the embryo remains dormant until the pouch is empty. After a pregnancy of 34–38 days the single young attaches firmly to one of 4 teats in the mother's pouch. It detaches from the teat at about 23 weeks, vacates the pouch at about 38 weeks and suckles at foot until about 15 months old. **Diet** Grass, herbs and ferns. They seldom need to drink. **Habitat** Open forest with a grassy understorey, often in hilly areas. **Traces** Scats are deposited in groups or clumped in strings. They are round to oval or square pellets, about 10–20 mm across and 20–40 mm long, containing coarse plant fragments. **Status** Low risk.

WEIGHT **7–26 kg** HEAD–BODY **675–955 mm** TAIL **725–1045 mm**

FAMILY **MACROPODIDAE** SPECIES *Macropus robustus*

COMMON WALLAROO OR EURO

The two major subspecies of this large kangaroo differ markedly in their colouration and habitat. The eastern subspecies is a dark grey forest-dweller known as the common wallaroo. It has fairly coarse, shaggy fur, and some females are bluish-grey, while some males have a reddish band across the shoulders and neck. In central and western Australia it has shorter reddish-brown fur, lives in hot, arid areas, and is known as the euro. All have a black naked nose, a short thick tail and relatively short limbs. Males attain twice the weight of females. **Behaviour** Common wallaroos shelter from the heat of the day among dense trees, in caves, or under rock ledges, often above their feeding grounds. They emerge in the late afternoon to graze, rarely venturing more than 200 m from a rock shelter. They are generally solitary and occupy small overlapping home ranges, including parts of the lower slopes and nearby plains. During courtship, males mark low vegetation with secretions from a chest gland and fight over females on heat, engaging in ritualised boxing matches to establish dominance and mating preferences. When alarmed they make threatening hisses, clucking calls and thump the ground with their hind feet. Like other wallaroos they stand with their wrists raised, shoulders thrown back and elbows close to their sides. **Development** Sexually mature at 18–24 months, common wallaroos may live to 15 years. They breed throughout the year and mate soon after giving birth, although the embryo remains dormant until the pouch is empty. After a pregnancy of about 34 days the newborn attaches firmly to one of 4 teats in the mother's pouch. It leaves the pouch at about 34–37 weeks and suckles at foot until about 14 months old. **Diet** Native grasses and some shrubs. They can survive for long periods without drinking. **Habitat** Rocky slopes, stony hills and escarpments; from wet sclerophyll forests to arid tussock grasslands; from the tropics to subalpine regions. **Traces** Water digs in dry river beds in desert areas. Scats are round, oval or square pellets; about 10–20 mm across and 20–40 mm long; containing fairly coarse plant fragments; and sometimes deposited in clumps. **Status** Low risk nationally, but vulnerable in Vic. and on Barrow Island, WA.

WEIGHT **18–46.5 kg** HEAD–BODY **570–1100 mm** TAIL **530–900 mm**

FAMILY **MACROPODIDAE** SPECIES *Macropus rufogriseus*

RED-NECKED WALLABY OR BENNETT'S WALLABY

This species is known as Bennett's wallaby in Tasmania and the Bass Strait Islands, where it is darker with a shaggier coat. The fur is soft and deep, dark brown to reddish-brown above with pale tan to white tips, and reddish-brown on the neck. The belly and chest are light grey; the upper lip has a white stripe; the nose, paws and longest toe are black. Females are paler. The tail is well-furred with a brushy tip. The hind feet have no first digit, the second and third are fused with a double claw, and the fourth is much longer than the others. Males are much larger than females. **Behaviour** Red-necked wallabies sleep by day in dense vegetation, emerging in the late afternoon (or earlier on cool days) to follow well-defined runways to feeding sites where they often congregate in groups of 30 or more. Males are generally solitary and occupy a home range of about 32 ha. Female home ranges are around 12 ha and overlap close female relatives. In the breeding season males range widely searching for females on heat, and establish dominance hierarchies by fighting and ritualised aggression. Subordinate males are excluded from mating by dominant males and females. Young females settle close to their mother, but males are forced out of their mother's home range at about 2 years of age. **Development** Red-necked wallabies have a lifespan of up to 18 years. Females are sexually mature at 11–21 months, males at 13–19 months. Mainland animals breed throughout the year, while those in Tasmania breed from January to July. Females mate soon after giving birth, but the embryo remains dormant until the pouch is empty (although in Tasmania the embryo remains dormant until the next breeding season). After a pregnancy of 29–30 days, 1–2 young are born and each attaches firmly to one of the 4 teats in the mother's pouch. Young leave the pouch at 40–43 weeks, and for the following month are left in a hiding place while the mother feeds. Thereafter they suckle at foot until 12–17 months. **Diet** Grasses and herbs. **Habitat** Wet and dry sclerophyll forests and woodlands, with grassy areas and a dense understorey; tall coastal heaths. **Traces** Well-worn runways to feeding areas. Scats are oval to round pellets of grass fragments, 10–20 mm across and up to about 30 mm long, deposited in groups or strings. **Status** Low risk nationally, but rare in SA.

WEIGHT **11–27 kg** HEAD–BODY **660–925 mm** TAIL **620–880 mm**

FAMILY **MACROPODIDAE** SPECIES *Macropus rufus*

RED KANGAROO

The red kangaroo is one of the largest living marsupials. Males are very powerfully built and much larger than females. The fur is reddish-brown above and paler below, although eastern females are blue-grey. A broad white stripe runs along the cheek, while the muzzle has black and white markings with a partially naked tip. The hind feet have no first digit, the second and third are fused with a double claw for grooming, and the fourth is much longer than the others. Males are larger than females. **Behaviour** Red kangaroos rest during the heat of the day in dusty scrapes under bushes or shrubs, and feed from late afternoon to early morning. On cool, wet days they may be active throughout the day. They live alone or in groups of up to 10, comprising mainly females and their offspring and one or more males, occupying a home range of around 8 sq km. This area increases in droughts when mobs of several hundred may gather to feed and drink. Males congregate around females on heat and establish dominance hierarchies by boxing or kicking. The largest males have exclusive mating rights and monopolise several females, driving other males away. Young males range widely and old males are solitary. When alarmed they make a loud cough and thump their hind feet, hopping rapidly away with the body and tail horizontal, reaching speeds of up to 64 kph. **Development** They have a lifespan of up to 22 years. Sexual maturity is reached at about 2 years, although in good years females can mate at 15 months and continue to breed throughout the year. Mating takes place soon after birth, although the embryo remains dormant until the pouch is empty. Pregnancy lasts 33 days and the newborn attaches firmly to one of 4 teats in the mother's pouch. It leaves the pouch at around 8 months and suckles at foot for another 3–4 months. In prolonged droughts more than half the pouch young die and females stop mating. **Diet** Native grasses, herbs and shrubs. It can survive without water if the food is green. **Habitat** Semi-arid plains, woodlands, grasslands and open forests. **Traces** Similar to the eastern grey kangaroo (page 103).
Status Low risk.

WEIGHT **17–85 kg** HEAD–BODY **740–1400 mm** TAIL **640–1000 mm**

FAMILY **MACROPODIDAE** SPECIES *Onychogalea fraenata*

BRIDLED NAILTAIL WALLABY

The bridled nailtail wallaby was widespread throughout central eastern Australia until the early 20th century, but was thought to be extinct until rediscovered in1973. It is now restricted to reserves in central Qld. This small wallaby is grey to light tan with a distinctive white 'bridle' stripe running from the centre of the back along the shoulders to behind the forearms. A black stripe runs along the spine, and there are white cheek stripes. Nailtail wallabies have a horny spur 3–6 mm long at the end of the tail, well-developed claws on the forefeet for digging, and move their forelimbs in a circular motion when hopping, hence the vernacular name 'organ-grinder'. Males are larger than females. **Behaviour** Bridled nailtail wallabies are nocturnal and generally solitary, although males sometimes rest with a female on heat and her young at foot. They live within a home range averaging around 60 ha for males and 25 ha for females, with an open nocturnal feeding area and an adjacent daytime resting area with plenty of cover, preferring brigalow regrowth habitats. They shelter by day in a hollow scraped beneath a tree or shrub, or in a hollow log, and use several shelter sites in their home range. At dusk they emerge to feed in grassy areas and return to cover at dawn. Male home ranges overlap those of 4–21 females, and males often gather around a female on heat, with the largest males defending access to the female. If disturbed they prefer to hide rather than flee. **Development** Males are sexually mature at 9–14 months and females at 5–9 months, although a male may not be large and strong enough to mate successfully until 18 months. They breed throughout the year with peaks in late spring and summer when food is plentiful, raising up to 3 young per year. A single young is born some 24 days after mating and attaches firmly to one of 4 teats in the mother's pouch. It leaves the pouch at about 4 months and then suckles at foot, becoming independent 7–8 weeks later. Females hide their young joeys close by in low, dense vegetation during the day. **Diet** Non-woody broad-leaved plants, succulents, flowering plants and grasses. **Habitat** Brigalow shrublands, open grassy eucalypt woodlands and dense acacia forests in semi-arid regions. **Traces** Scats are about 20 mm long and 10 mm wide, pointed at one end, light to dark brown, containing coarse plant fragments. **Status** Endangered. Threatened by land clearing, competition from stock and predation by introduced predators.

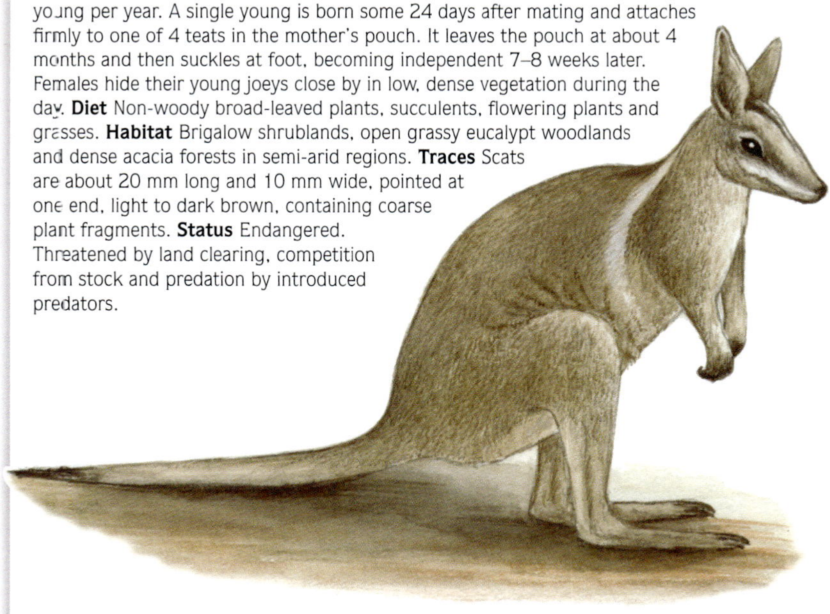

WEIGHT **4–8 kg** HEAD–BODY **430–700 mm** TAIL **360–540 mm**

FAMILY **MACROPODIDAE** SPECIES *Onychogalea unguifera*

NORTHERN NAILTAIL WALLABY

The northern nailtail wallaby derives its name from a small horny spur of unknown function hidden among a brush of dark hairs at the tip of its slender, whip-like tail. It is a small sandy-coloured wallaby with a dark stripe along the lower back and tail, and a faint dark shoulder stripe. It has a squarish muzzle, long ears and very slender upper incisor teeth inclined forward. The forefeet have long, well-developed claws used to dig shallow scrapes in the soil. The long hind feet have no first digit, the second and third are fused with a double claw for grooming, and the fourth is much longer than the others. Males are larger than females. **Behaviour** Active between dusk and dawn, northern nailtail wallabies rest by day in a shallow scrape in the shade under low trees, dense shrubs or grassy tussocks. They are usually solitary, but sometimes feed with up to 3 others, often travelling more than 3 km from daytime resting sites to feeding areas, always staying close to shelter. They move in a low crouching hop with the tail curved upwards, but when travelling fast they move their forelimbs in an unusual rotary action, once likened to that of an organ-grinder. **Development** They have a lifespan of up to 6 years, but little is known about their development. They probably breed year-round and females give birth to a single young which attaches firmly to one of 4 teats in her pouch. **Diet** Herbs, leaves, native grasses, fruits, stem bases and rhizomes. **Habitat** Open grassy woodlands, tall shrublands, tussock grasslands and shrubby savanna; usually near water. **Traces** Scats are oval pellets about 10 mm across and 15–25 mm long, containing coarse plant fragments. **Status** Low risk.

WEIGHT **4.5–9 kg** HEAD–BODY **490–690 mm** TAIL **600–730 mm**

FAMILY **MACROPODIDAE** SPECIES *Petrogale brachyotis*

SHORT-EARED ROCK-WALLABY

Although abundant and distributed over a wide area, little is known about the biology of the short-eared rock-wallaby. The fur is short and fine, varying from light grey above in western populations to dark grey and brown in eastern populations, and white to greyish-white below. It has a dark brown neck stripe, a white shoulder patch and a dark tip to the long, bushy tail. Arnhem Land individuals are dark grey to brown above with a distinct white or buff side stripe, a dark neck and back stripe. The ears are less than half the length of the head and have a whitish margin. The hind feet have no first digit, the second and third are fused with a double claw used for grooming, and the fourth is much longer than the others. The feet have thick, rough pads to give them a secure grip. Males are larger than females. **Behaviour** Short-eared rock-wallabies are mainly nocturnal, and sleep by day in caves, cool rocky crevices or among boulders. They are shy and wary, and emerge in the evening to feed in adjacent areas, often with others, and may bask in the sun on cool winter days. They maintain a home range of some 18 ha and use a range of rock piles and outcrops for dens, using on average 4 dens within outcrops and some more exposed resting sites. Individuals sometimes share dens, although males are generally intolerant of other males. **Development** They have a lifespan of about 10 years and probably breed throughout the year with a peak in the wet season, giving birth to a single young which attaches to one of the 4 teats in the mother's pouch. The young leave the pouch at about 30 weeks and continue suckling at foot for another 3 months. **Diet** Grass, leaves, herbs and seeds. **Habitat** Rocky sites and boulder-strewn outcrops including cliffs, low rocky hills and gorges; usually near woodlands, forests and savanna grasslands. **Status** Low risk.

WEIGHT **2.2–5.6 kg** HEAD–BODY **405–550 mm** TAIL **320–550 mm**

Family **Macropodidae** | Species *Petrogale lateralis*

Black-footed Rock-wallaby

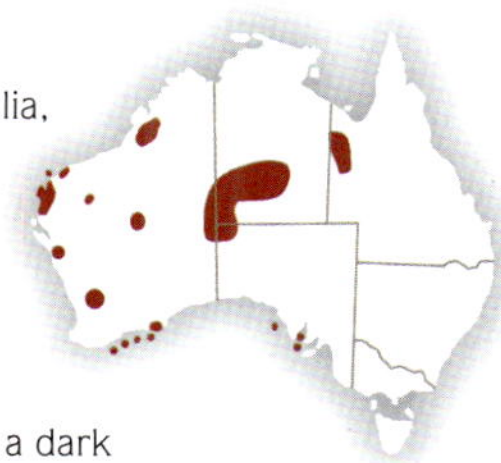

Found in arid and semi-arid areas of central and western Australia, the black-footed rock-wallaby comprises 4 subspecies and 2 geographic subpopulations, all differing in range and colour. It has thick, soft fur, dark to pale grey-brown above (sometimes purplish in western Queensland) with a paler chest and dark brown belly. A dark brown to black stripe runs between the ears to the mid-back. The cheeks sometimes have a white or sandy-coloured stripe. The feet are dark brown and the tail has a dark brown to black tip. The hind feet have no first digit, the second and third are fused with a double claw used for grooming, and the claw of the long fourth toe projects only slightly beyond the large pad. The soles of the feet are granulated to help them grip the rocks. **Behaviour** Black-footed rock-wallabies are shy, wary, and predominantly nocturnal, sheltering in caves and rocky crevices from the heat of the day to reduce water loss. They emerge in the early evening to feed in open areas, deriving most of their water needs from their food. On cool days they often bask in the sun. They are gregarious and live in groups of 10–100 individuals with separate male and female dominance hierarchies, established mainly by ritualised and aggressive acts, including kicking, but rarely fighting. They hop with great agility on rocky outcrops with the tail arched over the back. **Development** Females are sexually mature at 12–24 months. Breeding can be continuous and females mate again soon after giving birth. Development of the embryo is delayed until the pouch is vacated or, in dry years, until conditions are suitable for development. A single young is born some 30 days after mating and attaches firmly to one of 4 teats in the mother's pouch. After leaving the pouch, the young continues to suckle at foot and is often left in a safe place while the mother forages. **Diet** Grasses and other vegetable matter. They rarely need to drink. **Habitat** Rock piles, steep cliffs, boulder scree slopes and granite outcrops, typically with some open vegetation to provide cover. **Status** Vulnerable. Threatened by competition from rabbits and predation by feral animals.

Weight **2.3–7.1 kg** | Head–body **445–521 mm** | Tail **320–605 mm**

FAMILY **MACROPODIDAE** SPECIES *Petrogale penicillata*

BRUSH-TAILED ROCK-WALLABY

The brush-tailed rock-wallaby has a dull-brown back, a rufous rump, black furry feet and a long, densely-furred tail with a bushy tip. Southern animals are more colourful with a black tail and armpits, pale stripes along the sides, white cheek stripes and a black stripe on the forehead. Their ears are yellowish inside, with black patches and whitish margins. The hind feet have no first digit, the second and third are fused with a double claw used for grooming, and the claw of the long fourth toe projects only slightly beyond the large pad. The pads are granulated for gripping, with a fringe of stiff hairs. Males are larger than females. **Behaviour** Brush-tailed rock-wallabies are mainly nocturnal, sheltering by day in caves, deep rock fissures and dense stands of lantana, sometimes basking in the winter sun. They form small colonies with dominance hierarchies, although individuals have overlapping home ranges of around 15 ha with exclusive den sites. Colonies often comprise a single male and 1–3 females and their young. Females are gregarious and frequently share den sites with female relatives and groom each other regularly. They hop with great agility on rocky outcrops with the tail arched over the back, and can ascend almost vertical rock faces and climb sloping tree trunks. Young at foot are left in the den while the mother looks for food or drink, often travelling 400 m or more from her daytime shelter to nocturnal foraging areas. **Development** Females are sexually mature by 18 months, males by 20 months, and have a lifespan of 14 years in captivity. They breed all year, mating soon after giving birth, although development of the embryo is delayed until the pouch is vacated. After a pregnancy of 31 days a single young is born and attaches to one of 4 teats in the pouch, where it remains for about 29 weeks, thereafter suckling at foot for about 3 months. **Diet** Mainly grasses, supplemented with leaves, sedges, ferns, roots, bark, fruit, seeds and flowers. **Habitat** Rocky areas with large tumbled boulders, ledges and caves; close to grassy sites in a range of habitats including rainforests, sclerophyll forests and open woodlands. **Traces** Scats are found in groups of 4–8 near rock shelters and feeding sites. They are cylindrical pellets, 10 mm or more across and 15–30 mm long, usually brown with coarse plant fragments. **Status** Endangered in NSW and Vic.; vulnerable in Qld. Threatened by fox predation and competition from goats, sheep and rabbits.

WEIGHT **4.9–10.9 kg** HEAD–BODY **450–586 mm** TAIL **500–700 mm**

FAMILY **MACROPODIDAE** SPECIES *Petrogale xanthopus*

YELLOW-FOOTED ROCK-WALLABY

A very attractively ornamented and brightly-coloured species, the yellow-footed rock-wallaby is greyish-fawn above with rich brown stripes along the centre of the back, hips and arms, white stripes along the cheeks and sides, rufous-brown arms and legs, and white underparts. The tail is usually rufous-brown with dark bands, although the pattern varies and the tip may be white. The ears are long and the nose naked. The hind feet have no first digit, the second and third are fused with a double claw used for grooming, and the fourth is much longer than the others. The claw of the fourth toe projects only slightly beyond the large pad. The pads are granulated for gripping, with a fringe of stiff hairs around them. Males are larger than females. **Behaviour** Active mainly at night, yellow-footed rock-wallabies shelter by day under rocky outcrops or among vegetation between boulders, emerging in the evening to feed. In winter they sometimes bask in the sun. They are gregarious, forming colonies of up to 100 individuals each with overlapping home ranges of 150–200 ha. Each rockpile within the colony's boundary is occupied by a group of females, a large male and several smaller males. Males establish dominance hierarchies by ritualised aggressive acts and occasionally fighting. They are agile, hopping among rocks with the tail arched over the back, and are able to climb sloping tree trunks. The young are left behind in a safe place while the mother forages for food or drink, often travelling long distances. The young make a clicking call when lost or disturbed. **Development** Females reach sexual maturity between 11 and 22 months, males at about 30 months, and they have a lifespan of 10 years or more. They breed throughout the year with peaks in April and September and mate soon after giving birth, although development of the embryo is delayed until the pouch is vacated. After a pregnancy of 31–32 days a single young is born and attaches to one of the 4 teats in the mother's pouch, where it remains for about 28 weeks, thereafter suckling at foot until weaned. **Diet** Grass, leaves, herbs, tubers and bark. They need access to water and drink frequently in hot weather. **Habitat** Semi-arid rocky hills and outcrops, usually with mulga scrub. **Traces** Scats are deposited in groups around rockpiles. They are cylindrical to oval, 10–15 mm across and 20–30 mm long, comprising coarse plant fragments. **Status** Low risk in Qld, but endangered in NSW and SA. Threatened by competition and predation by feral animals.

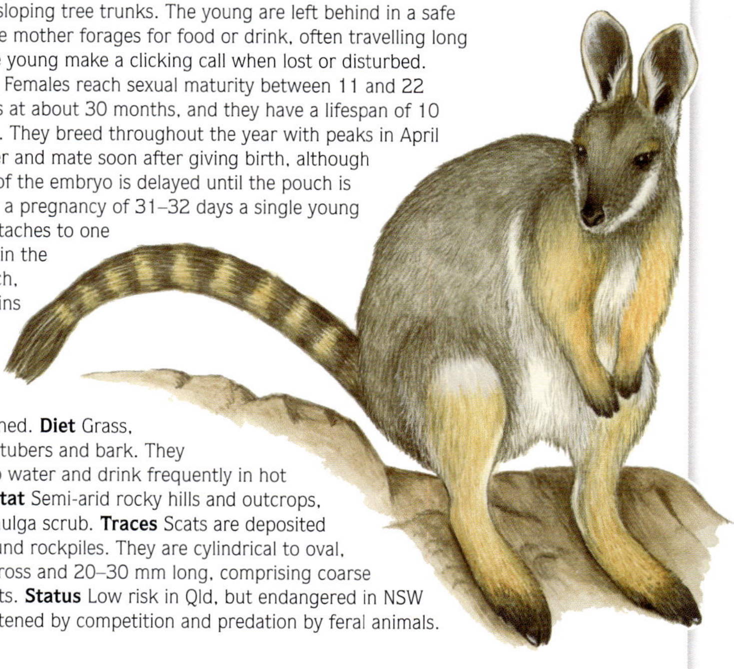

WEIGHT **6–11kg** HEAD–BODY **480–650 mm** TAIL **560–700 mm**

FAMILY **MACROPODIDAE** SPECIES *Thylogale billardierii*

TASMANIAN PADEMELON

A small stocky macropod, the Tasmanian pademelon has long, dense, soft fur. It is dark brown to golden-brown above with a grey base, and pale reddish-buff below with a grey to white base. The tail is short and thick, the ears rounded and the nose naked. The hind feet have no first digit, the second and third are fused with a double claw used for grooming, and the fourth is much longer than the others. Males are much larger than females. **Behaviour** Tasmanian pademelons are generally active from dusk to dawn and rarely venture more than 100 m from the forest edge. They sleep by day in dense cover, although in winter and when the weather is bad they emerge to feed during the day, digging food from the snow in alpine areas. They live alone in a home range as large as 170 ha, and travel along well-established runways to communal feeding sites up to 2 km away where they congregate in groups of 10 or more. They have a short rapid hop and escape to dense undergrowth when alarmed. Males establish dominance hierarchies during the breeding season by fighting and ritualised aggression. Top-ranking males have exclusive mating rights with females. Males utter a guttural growl or hiss when fighting and make clucking sounds during courtship. **Development** Tasmanian pademelons are sexually mature at about 14 months and have a lifespan of up to 9 years in captivity. Although they are capable of breeding year round, most births occur from April to June. Females mate within 24 hours of giving birth, but the embryo remains dormant until the pouch is empty. After a pregnancy of 30 days a single young is born and attaches to one of the 4 teats in the mother's pouch. The infant leaves the pouch at about 29 weeks and suckles at foot until about 10 months old. **Diet** Soft grasses, herbs, leaves and seedlings, often held in the forepaws. **Habitat** Coastal and montane forests and scrubs with dense undergrowth and adjacent grassy areas. **Traces** Small resting hollows in dense vegetation and well-worn runways. Scats are oval to cylindrical, about 10–15 mm across and 15–25 mm long, brown, containing coarse plant fragments. **Status** Low risk.

WEIGHT **2.4–12 kg** HEAD–BODY **360–720 mm** TAIL **320–485 mm**

FAMILY **MACROPODIDAE** SPECIES *Thylogale stigmatica*

RED-LEGGED PADEMELON

A small, compact macropod generally found in rainforests and other densely-vegetated habitats, the red-legged pademelon has thick soft fur, grey-brown to dark brown above with red-brown markings on the cheeks, thighs and forearms. The ears are rounded, the nose naked and the tail short and thick. The hind feet have no first digit, the second and third are fused with a double claw used for grooming, and the fourth is much longer than the others. Males are larger than females. **Behaviour** Red-legged pademelons are active most of the day and night, although they usually rest between midday and early afternoon and around midnight. They sleep in refuges in dense cover with the head on the tail or ground, and rest sitting with the tail between the legs, leaning back on a rock or tree. Daylight hours are spent searching for food within the forest, foraging on the ground and moving slowly on all fours. After dusk they follow well-defined runways through the vegetation to feeding sites on the forest edge. Shy and generally solitary, they have home ranges of 1–4 ha, and rarely venture more than 70 m from the forest edge. They sometimes feed in small groups and when disturbed they thump the ground with the hind feet as they dash for cover. Vocalisations include a soft clucking by females calling their young and by males during courtship; and a harsh rasp made by females in hostile situations or when rejecting courtship advances by a male. **Development** Males and females become sexually mature at about 66 weeks and 48 weeks respectively. Females give birth to a single young 28–30 days after mating, and mate again 2–12 hours later, although development of the embryo is delayed until the pouch is empty. The newborn attaches to one of 4 teats in the mother's well-developed pouch, detaching at 13–18 weeks, opening its eyes at 16–18 weeks and making short excursions from the pouch at 22–26 weeks. Young vacate the pouch at 26–28 weeks and are weaned 9–10 weeks later. **Diet** Leaves, fruit, ferns, fungi and grasses. **Habitat** Rainforests, wet sclerophyll forests and vine scrubs. **Traces** Small resting hollows and well-worn runways in dense vegetation. Scats are about 10 mm across and 20 mm long, black or brown, containing coarse plant fragments. **Status** Low risk in Qld, but vulnerable in NSW. Threatened by habitat destruction and predation by feral animals.

WEIGHT **2.5–6.8 kg** HEAD-BODY **385–540 mm** TAIL **300–475 mm**

FAMILY **MACROPODIDAE** SPECIES *Thylogale thetis*

RED-NECKED PADEMELON

The red-necked pademelon is a compact macropod of the forest edge, distinguished from the red-legged pademelon by the reddish-brown fur around its neck and shoulders. The fur is thick and soft, brownish-grey above and whitish below. The ears are rounded, the nose naked and the tail short and thick. The hind feet have no first digit, the second and third are fused with a double claw used for grooming, and the fourth is much longer than the others. Males are larger than females. **Behaviour** Active most of the day, red-necked pademelons move slowly through the forest in daylight hours between bouts of sleep, seeking food, and basking in the sun in winter. After dusk they travel along well-defined runways to forest clearings and feeding sites at the forest edge, returning to cover just before dawn. They sleep in dense vegetation in shallow depressions in the leaf litter. Shy and generally solitary, they have home ranges of 5–30 ha and rarely venture more than 100 m from the forest edge. Small feeding groups often congregate in clearings at night, hopping quickly back to cover if disturbed with the tail held out stiffly behind. Slow movements are quadrupedal with the tail dragging behind. Vocalisations include a threatening growl, a repetitive click by females calling their young and a clucking sound made by males during courtship. If alarmed they thump their hind feet as they hop away. Males establish dominance hierarchies by fighting or ritualised aggression. **Development** Sexually mature at 17 months, they breed all year with birth peaks in autumn and spring in northern areas and summer in the south. Females can become pregnant soon after giving birth, but development of the embryo is delayed until the pouch is empty. A single young is born and attaches to one of the 4 teats in the mother's pouch, vacating the pouch by 26 weeks and suckling at foot for a further 4 weeks. **Diet** Grass, herbs, leaves, and bark; often holding food in the forepaws. **Habitat** Margins of closed forests and rainforests. **Traces** Similar to the red-legged pademelon (page 117). **Status** Low risk.

WEIGHT **3.8–9.2 kg** HEAD-BODY **290–620 mm** TAIL **270–510 mm**

Family **Macropodidae** Species *Setonix brachyurus*

Quokka

The quokka is a small, robust wallaby with long, thick, coarse fur flecked with grey and brown, tinged rufous on the back and pale grey below. The head is broad with a dark stripe on the forehead and small rounded ears set on top. The tail is thick and sparsely-haired with visible scales. The hind feet have no first digit, the second and third are fused with a double claw used for grooming, and the fourth is much longer than the others. **Behaviour** Active mainly at night, quokkas shelter by day in groups in dense vegetation and converge at night around waterholes within a group territory occupied and defended by 25–150 adults. In dry conditions adults sometimes fight for available shelter and water. They live in overlapping home ranges of around 6 ha and adult males establish dominance hierarchies according to age. Females and juveniles have no rank in the hierarchy. Males defend areas around their resting sites and may form long term bonds with females. They move with a bounding gait interspersed with short bouts of rapid hopping, and may climb trees to reach twigs up to 2 m above ground. **Development** Quokkas have a lifespan of 13 years or more. Females are sexually mature by about 9 months and males by about 13 months. They breed throughout the year on the mainland, and from January to March on Rottnest Island. Females mate again soon after giving birth, but development of the embryo is delayed until the pouch is empty and will resume if conditions are favourable. After a pregnancy of 26–28 days a single young is born and attaches to one of 4 teats in the mother's pouch. The young vacates the pouch by 26 weeks and suckles at foot for 2 months. **Diet** Grasses, leaves and succulents. They can survive for long periods without water and have been observed drinking seawater. **Habitat** Wet and dry sclerophyll forests, woodlands and heath. **Traces** Scats are found in groups at feeding sites and resting places. They are oval to oblong, about 15 mm across and 15–25 mm long, containing coarse plant fragments. **Status** Vulnerable. Threatened by land clearing and fox predation.

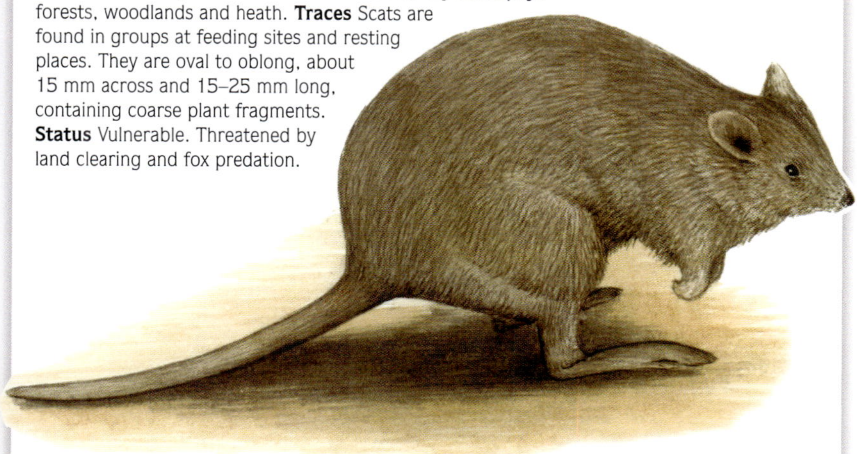

Weight **2.7–4.2 kg** Head–body **400–540 mm** Tail **245–310 mm**

FAMILY **MACROPODIDAE** SPECIES *Wallabia bicolor*

SWAMP WALLABY

A common inhabitant of the dense forest understorey, the swamp wallaby has coarse fur, dark brown to black flecked with yellow above, and red-brown to yellow-brown below. The cheeks and shoulders have a light yellow to red-brown or black stripe; the paws, feet and occasionally the end of the tail are very dark brown. The hind feet have no first digit, the second and third are fused with a double claw for grooming, and the fourth is much longer than the others. Males are larger than females. **Behaviour** Swamp wallabies are generally active from dusk to dawn, resting in thick undergrowth during the day and moving out to more open grassy areas around the forest edge at night to feed. In winter they often forage during the day. Essentially solitary, swamp wallabies have home ranges of 2–16 ha, and sometimes feed in small groups. They seem relatively uncoordinated when moving slowly, holding the tail up high. In long grass they often make high leaps. When moving fast they hold the head low and the tail horizontal. **Development** Sexually mature at 15–18 months, swamp wallabies may live to 15 years in the wild. When dingo predation is high they breed year round, but when predation is low they are more seasonal, with most births occurring between April and November. Females mate again about 7 days before giving birth, producing a single embryo that remains dormant until the pouch is empty. After a pregnancy of 33–38 days the newborn attaches firmly to one of the 4 teats in the mother's pouch, which it vacates at about 36 weeks, thereafter suckling at foot until about 15 months old. **Diet** Shrubs, seedlings, rushes, fungi, vines, ferns and grasses. Food is manipulated with the forelimbs. **Habitat** Rainforests, sclerophyll forests and woodlands with a dense understorey; heathlands; to 1200 m. **Traces** Scats are oval to round pellets 20 mm or more across, deposited in groups, or joined into cylinders when fresh growth is eaten in spring. They are usually brown when broken, comprising coarse plant fragments. **Status** Low risk nationally, but endangered in SA. Threatened by habitat destruction and predation by feral animals.

WEIGHT **10–21 kg** HEAD–BODY **665–847 mm** TAIL **640–862 mm**

FAMILY **NOTORYCTIDAE** SPECIES *Notoryctes typhlops*

SOUTHERN MARSUPIAL MOLE

This elusive, fascinating animal closely resembles the eutherian golden mole of Africa. Adapted to an underground existence, the marsupial mole is blind with tiny vestigial eyes, and lacks external ears, having only small holes covered by dense hair. A horny shield protects the nostrils, while the tail is reduced to a leathery stub. The limbs are short and powerful. The third and fourth toes of the forefeet are modified for digging and have large triangular claws forming a cleft spade or scoop. The remaining 3 toes are small, the first and second are opposed to the third and fourth toes. The middle 3 toes of the hind feet also have enlarged claws. The body is covered with long, silky, golden-brown fur. **Behaviour** Marsupial moles live mostly underground, only occasionally venturing to the surface, usually after rain. They are extensive burrowers, digging with the flattened claws of the forefeet, pushing the soil out with the hind feet and tail while they progress forward. Most burrows are 200–1000 mm below the surface. Although some are more than 2 m deep. The burrows comprise horizontal tunnels and occasional vertical shafts and are backfilled as the mole moves along. Marsupial moles require soft sand and cannot tunnel through hard and loamy ground. They sleep below ground and occasionally come to the surface, particularly after rain in cooler seasons. On the surface they pull themselves along in a rapid shuffle, changing direction frequently, leaving 3 parallel furrows behind made by the feet and stumpy tail. They only stay on the surface for short distances of a few metres. Marsupial moles make sharp squeaking sounds if disturbed and are thought to be solitary. **Development** Males lack a visible scrotum: the testes are located between the skin and abdominal wall. Females have a deep, rear-opening pouch completely covering 2 teats. Little is known about their reproduction and development, although single and twin pouch young have been reported. **Diet** Ants, termites, beetles, larvae and seeds. **Habitat** Sandy deserts, occupying dunes, sand plains and river flats. **Traces** Tracks of 3 parallel furrows in the sand, leading to a small hole. **Status** Endangered, with a population size estimated to be fewer than 10,000 adults.

WEIGHT **40–70 g** HEAD–BODY **120–160 mm** TAIL **20–26 mm**

FAMILY **PTEROPODIDAE** SPECIES *Macroglossus minimus*

NORTHERN BLOSSOM-BAT

An important pollinator of paperbark trees, the northern blossom-bat is one of the smallest fruit bats and has a wingspan of 130–150 mm. It probes flowers for nectar and pollen with its long, narrow, brush-tipped tongue and is similar to the common blossom-bat with long soft fur, fawn to reddish-brown above and paler on the chest. It can be distinguished by the presence of flap-like membranes ('pyjamas') on each leg. It has a small rudimentary tail and a fox-like head with a tapering muzzle and simple, broad, rounded ears. The first finger has a large claw and the second finger a tiny claw. The incisor teeth are slender, weak, and often malformed. **Behaviour** Northern blossom-bats roost alone in thick foliage, under loose bark, in tree hollows, rolled banana leaves and under roofs, hanging upside down, wrapped in their wings. They can be active by day or night, and navigate by sight, flying slowly and foraging in forests and woodlands for flowering and fruiting trees, often detected by smell. They hover beside a flower, or land on it, and lap up the nectar with their long, bristly tongue. Pollen sticks to the fur and is licked off during a grooming session. Individuals chase other bats away from their feeding trees, screeching and slapping their wings. Adult males have a pink, V-shaped gland on the chest, producing a pungent smell which probably attracts females. **Development** Males become sexually mature at 7 months, females at about 10 months. Breeding may occur at any time of year, and births have been recorded throughout the year with peaks in March–April and August–September. A single young is born about 4 months after mating and suckles from a teat under the mother's armpit, clinging constantly to her fur with its clawed feet for 6–10 days. It then remains at the roost while she forages, can fly by 40 days and is weaned at 60–70 days. **Diet** Nectar, pollen and rainforest fruit, particularly figs. **Habitat** Rainforests, mangroves, paperbark woodlands, banana plantations and monsoon scrub; from sea level to 1500 m. **Status** Low risk.

WEIGHT **10–20 g** HEAD–BODY **49–67 mm** FOREARM **37–43 mm**

FAMILY **PTEROPODIDAE** SPECIES *Syconycteris australis*

COMMON BLOSSOM-BAT

Also known as the Queensland blossom-bat, this mouse-sized species is an important pollinator in the tropics, feeding on nectar obtained with its long, thin, brush-tipped tongue and slim, pointed muzzle. It has long, soft, reddish-brown to fawn fur extending to the ankle. This bat lacks a tail and has a fox-like head with simple rounded ears and obvious upper incisor teeth that are weak and slender. It has a large claw on the first finger and a tiny claw on the second finger. **Behaviour** Common blossom-bats roost by day alone in dense foliage, usually in the sub-canopy layer, hanging upside down with the wings wrapped around the body. They navigate by sight, flying 3–5 m above the ground, often following tracks through the forest to feeding sites up to 4 km away. They hover over selected blossoms and crash-land to feed, gathering nectar with the tongue. Pollen lodges on scale-like projections on the fur and is collected by grooming after a bout of feeding. Individuals defend food sources from intruders, chasing them away, shrieking and clapping their wingtips together. They live in overlapping home ranges of around 5.5 ha, changing roosts frequently within a 50 m radius. They move to the warmer edge of the rainforest in winter, becoming torpid for up to 12 hours on cold days or when food is scarce. **Development** Females give birth to a single young in October–November and another between February and April, although in tropical areas they may breed throughout the year. The young suckles from a teat in the mother's armpit for up to 3 months and is carried with her until well-furred. **Diet** Nectar, pollen and fruit, supplemented by some leaf. **Habitat** Littoral rainforests. **Status** Low risk in Qld, but vulnerable in NSW. Threatened by habitat destruction.

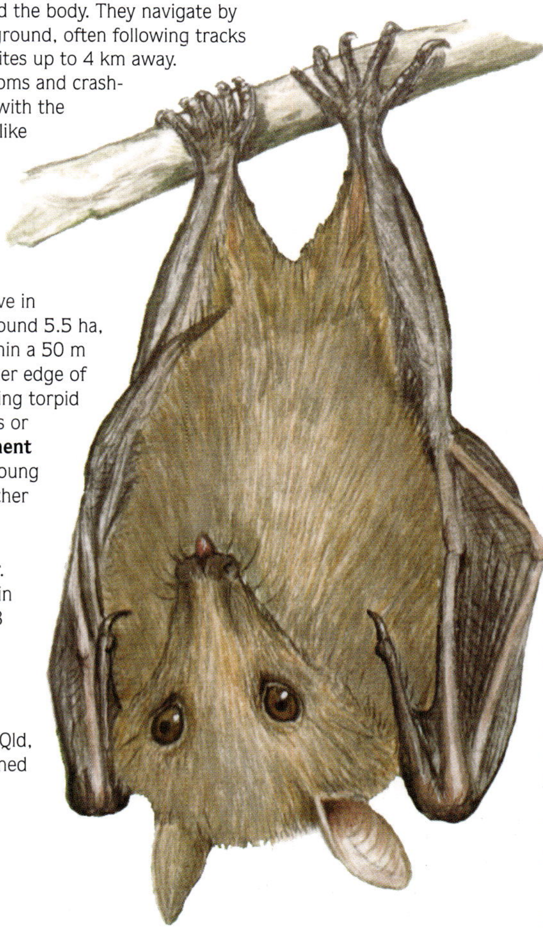

WEIGHT **13–26 g** HEAD–BODY **40–71 mm** FOREARM **38–45 mm**

FAMILY **PTEROPODIDAE** SPECIES *Nyctimene robinsoni*

EASTERN TUBE-NOSED BAT

Also known as the Queensland tube-nosed bat, this distinctive species is characterised by its prominent tubular nostrils that extend 5–6 mm beyond the muzzle, and by the pale yellow-green spots on its wings and ears. The fur is long and soft, grey-brown to light brown above and paler below, usually with a dark brown stripe along the centre of the back. It has simple rounded ears, a short, free tail and claws on both first and second fingers. **Behaviour** Eastern tube-nosed bats roost by day alone in trees with dense foliage, usually within about 200 m of their feeding sites. They hang from a branch, wrapped in their wings, and are very well camouflaged, closely resembling a dead leaf. Some bats remain in the same area for 3 years or more, while others are nomadic and may change roosts daily to take advantage of fruiting trees. Large feeding groups sometimes form around trees and individuals will aggressively defend a feeding territory. They have good vision and avoid flying in total darkness. They forage in the forest understorey, flying close to the ground along forest tracks, hovering with ease, often making high-pitched bleating calls. **Development** Females reach sexual maturity at 7 months. Breeding takes place in winter, and females give birth to a single young between October and December, some 4–5 months after mating. The newborn suckles from a teat in the mother's armpit and is carried with her until well-furred. **Diet** Rainforest fruits (particularly figs, blue quandongs and burdekin plums) and banksia blossoms. They can survive without drinking. **Habitat** Tropical and subtropical rainforests; wet sclerophyll forests with a well-developed rainforest understorey; usually beside streams. **Status** Low risk in Qld, but vulnerable in NSW. Threatened by habitat destruction and fragmentation of available habitat.

WEIGHT **30–56 g** HEAD–BODY **75–130 mm** FOREARM **60–70 mm** TAIL **20–25 mm**

Family **Pteropodidae** | Species *Pteropus alecto*

Black Flying-fox

The black flying-fox is a large coastal bat with a wingspan over 1 m; short, soft, black fur, often flecked with grey below, sometimes reddish-brown on the shoulders and back of the neck. It lacks a tail and has a fox-like head with large eyes and simple ears. The toes and first and second fingers have long, curved claws. **Behaviour** Black flying-foxes roost by day in the branches of large trees, hanging upside down wrapped in their wings. In spring they congregate in camps of several hundred to hundreds of thousands of bats, often shared with grey-headed and little red flying-foxes. They leave for nocturnal feeding grounds at dusk and fly with rapid wing beats at 35–40 kph, travelling up to 50 km, using sight and smell to find food. Old males usually live on the camp periphery and guard the camp from intruders. In the breeding season large males establish territories about 1 m across along branches marked by secretions from their shoulder glands. Their testes descend into a prominent position and they spend much time grooming and displaying their genitalia to attract females. At the end of the breeding season most bats disperse to live alone or in small groups. **Development** Mating takes place in March and April and a single young is born from August to November (January to March in the north) and suckles from a teat in the mother's armpit. It is carried for 3–4 weeks until well-furred, clinging to its mother's fur with its feet. It is then left in the camp at night. It can fly at 2 months and is independent at about 13 weeks. **Diet** Pollen and nectar of mostly eucalypt and paperbark trees; rainforest fruit and leaves. They can drink seawater. **Habitat** Rainforests, mangroves, wet sclerophyll forests and woodlands. **Traces** Similar to the grey-headed flying-fox (page 127). **Status** Low risk nationally, but vulnerable in NSW. Threatened by habitat destruction, shooting, and harassment at roosts.

Weight **500–880 g** | Head–body **186–300 mm** | Forearm **138–191 mm**

FAMILY **PTEROPODIDAE** SPECIES *Pteropus conspicillatus*

SPECTACLED FLYING-FOX

An essential pollinator and distributor of rainforest seeds, the spectacled flying-fox is named for the patches of pale yellow fur around the eyes and along the muzzle. It has a prominent yellow or silver-blond collar partially encircling the neck and soft black body fur often flecked with white. It has a fox-like head with large eyes and simple ears, and has no tail. The 5 toes and first and second fingers bear long curved claws. **Behaviour** Spectacled flying-foxes roost by day hanging from the branches of large trees, wrapped in their wings. They navigate by sight and fly off at dusk to feed in flowering and fruiting trees 20 km or more from their camp. Well-defined territories are established in feed trees, and intruders are chased away with much shrieking and wing-slapping, although they usually leave with a fruit. They form more or less permanent camps, some comprising many thousands of bats, in coastal sites in or close to the rainforest. During the mating season males attract females and mark territories by smearing their neck fur and branches with a red fluid containing secretions from sebaceous glands on the neck, mixed with a strong-smelling liquid secreted from the erect penis. Pregnant females leave the coastal camps and form maternity camps in highland regions. **Development** Sexually mature at 2 years, they mate from March to May and give birth to a single young from October to December. Newborn suckle from a teat in the mother's armpit and are carried for 1–2 weeks until well-furred. They are then left in the camp at night, are weaned at about 12 weeks and become independent at 5 months. **Diet** Rainforest fruit, pollen and nectar. They drink while skimming over the surface of water in flight and will drink seawater. **Habitat** Rainforests, mangroves, paperbark swamps and wet sclerophyll forests. **Traces** Similar to the grey-headed flying-fox (page 127). **Status** Vulnerable. Threatened by habitat destruction and shooting.

WEIGHT **500–850 g** HEAD–BODY **220–240 mm** FOREARM **155–181 mm**

FAMILY **PTEROPODIDAE** SPECIES *Pteropus poliocephalus*

GREY-HEADED FLYING-FOX

The largest Australian bat with a wingspan of up to 1.3 m, the grey-headed flying-fox has long fur, dark brown above and grey below, often flecked with white or yellow-brown. It has a light grey, fox-like head with a reddish-brown mantle encircling the neck, and has no tail. Thick leg fur extends to the ankle. The eyes are large and the ears simple. Both first and second fingers of the forelimbs are clawed and used for climbing. **Behaviour** Grey-headed flying-foxes navigate by sight and avoid flying in complete darkness. They leave the roost at dusk and fly to foraging areas up to 50 km away, travelling at 35 kph or more, and returning before dawn. They congregate in large camps, sometimes with tens of thousands of individuals, usually in gullies near water with good tree cover. They rest and sleep hanging from branches wrapped in their wings. Pregnant females form maternity camps before giving birth in September and October, and are joined by adult males as the young are being weaned from February to March. Breeding males mark territories and attract females by smearing secretions from their shoulder glands on their neck fur and branches. They defend a small area by fighting, wing-slapping and shrieking. Juveniles and old males keep guard on the camp perimeter. Adults usually disperse in winter, travelling up to 750 km, alone or in small groups. Juveniles and a few adults form winter camps near reliable food sources. **Development** Sexual maturity is reached at 18 months, although males first breed at about 30 months. Mating may take place almost any time of year, but most conceptions occur in March and April. A single young is born 6 months later and suckles from a teat in the mother's armpit while clinging to her belly fur. Young are carried on foraging flights for 3–4 weeks, until well-furred, and are then left in a crèche at the camp at night. They can fly at 3–4 months and are weaned at 5–6 months.

Diet Pollen, nectar and fruit (mainly figs). They can drink seawater.

Habitat Wet and dry sclerophyll forests and woodlands, sub-tropical and temperate rainforests, heaths, paperbark swamps and mangroves; to 700 m.

Traces Fibrous remnants of chewed fruit and flower clusters beneath trees. Scats are pungent, shapeless masses, beneath roosting and feeding trees.

Status Vulnerable. Threatened by habitat destruction, disturbance at roosting sites, shooting, and electrocution on power lines.

WEIGHT **600–1050 g** HEAD–BODY **230–290 mm** FOREARM **138–180 mm**

FAMILY **PTEROPODIDAE** SPECIES *Pteropus scapulatus*

LITTLE RED FLYING-FOX

The little red flying-fox is the smallest of the Australian flying-foxes, with a wingspan of about 1 m. This winged placental mammal has soft, short, dense fur, reddish-brown to yellowish-brown above and paler below with a light brown to yellow collar encircling the neck, patches of creamy-brown where the wings meet the shoulders, pale yellow hairs on the underside of the translucent wing membranes, and sparsely furred or naked legs. The fox-like head has large eyes and simple ears. It has no tail and the first 2 fingers of the forelimbs have claws.

Behaviour Active at night and on overcast afternoons, little red flying-foxes are highly nomadic and travel long distances using their keen senses of sight and smell to detect food. They are not very manoeuvrable and often crash-land into trees. In October–November they congregate in camps of up to a million individuals. Hundreds may share the same tree, hanging close together by their feet, wrapped in their wings, often breaking branches with their weight. Columns of bats leave at dusk, spiralling around before heading to feeding sites up to 50 km away. Juveniles live on the periphery of the camps while adult males gather harems of 2–5 females in small territories, marked by secretions of their shoulder glands, and defended from other males. After mating, around February, they tend to move to new camps in small, single sex groups. They migrate south in late spring/early summer and return north in late summer/autumn. **Development** Sexually mature at 18 months, they have a lifespan of up to 16 years. They mate around November–December and a single young is born around May–June. The newborn suckles from a teat in the mother's armpit and clings to her fur while she flies until well-furred at about one month. The young are then left in a crèche at the roost while she forages at night until 2 months old when they can fly. They are cared for until several months old when they have learned to forage and navigate. **Diet** Nectar and pollen of eucalypt and melaleuca trees, and rainforest fruits. **Habitat** Rainforests, mangroves, sclerophyll forests, paperbark swamps and woodlands. **Traces** Similar to the grey-headed flying-fox (page 127).
Status Low risk.

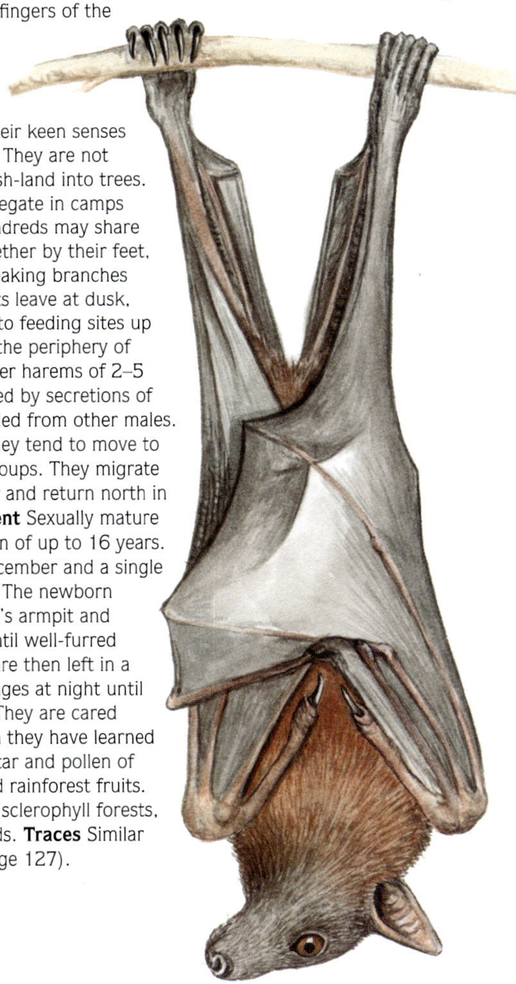

WEIGHT **310–605 g** HEAD-BODY **190–240 mm** FOREARM **120–155 mm**

Family **Megadermatidae** Species *Macroderma gigas*

Ghost Bat

Also known as the false vampire bat, this large predatory bat has a wingspan of about 500 mm, long soft fur, fawn-grey to dove-grey above and pale grey to white below. The ears are very large, joined above the head and sparsely-haired with a forked lobe partially covering the aperture. They have no tail. The eyes are large and a large simple noseleaf projects above the snout. **Behaviour** Active mainly at night, ghost bats roost by day in deep caves, rock fissures and mines, hanging from the ceiling alone or in small groups at least 250 mm apart and 2 m above the floor. Colonies of up to 1500 form in the breeding season. Mothers wean their young on prey brought back to the roost, and take juveniles with them to hunt. They fly fast and direct with the head held high, orienting by sight and by the intermittent use of echo-location, emitting signals through the mouth, foraging in preferred sites about 2 km from their roost. They often locate prey by sight and sound while sitting on a branch, and swoop on small terrestrial animals, enveloping them in their wings and biting them to death. Other bats are sometimes caught in flight. Prey is consumed at an established feeding site, usually in a small cave or rock overhang. They make twittering and cricket-like calls.

Development Ghost bats have a lifespan of up to 23 years. Mating takes place in April and May, and a single young is born in July and August. The newborn weighs 20 g and suckles from a teat in the mother's armpit for several months. It is carried until well-furred at 4 weeks, and then left in the roost while the mother hunts, until able to fly by 7 weeks. Young are weaned at about 3 months. **Diet** A variety of animals including large insects, birds, lizards, frogs, mice and other bats.
Habitat Rainforests, wet and dry sclerophyll forests and arid woodlands.
Traces Accumulations of bones and other animal remains at feeding sites.
Status Vulnerable. Threatened by mining of their roosting caves.

Weight **130–170 g** Head–body **98–130 mm** Forearm **96–118 mm**

FAMILY **RHINOLOPHIDAE** SPECIES *Rhinolophus megaphyllus*

EASTERN HORSESHOE BAT

The eastern horseshoe bat is a small, insectivorous bat with long, soft, greyish-brown fur, although some northern animals are orange-brown. It has large, simple and sharply-pointed ears, very small eyes and a complex noseleaf above the snout with a horseshoe-shaped lower part, a scalloped upper section and a fleshy central projection. The tail does not protrude beyond the tail membrane. **Behaviour** Active mainly at night, eastern horseshoe bats roost by day in warm, humid sites in caves, mines and buildings, forming colonies of up to around 50 bats outside the breeding season. They hang from the roof, 150–200 mm apart, with their feet together and head down. They travel up to 2 km from their roost sites, foraging for flying insects within or at the edge of the forest or woodland, flying slowly, changing direction suddenly and hovering. They occasionally hunt from a perch and often eat insects at the hunting perch. They have good eyesight but use echo-location to navigate and locate insects in the dark, emitting high-pitched sounds through the nose and directing these sounds with the noseleaf. Females congregate in maternity colonies of up to 4000 bats in spring, choosing warm humid caves to rear their young. Large colonies of up to 2000 non-breeding bats sometimes gather in the tropical dry season, while in temperate areas they disperse in winter and roost alone in cooler parts of their caves, becoming torpid during the day or hibernating from April to June in cold climates. **Development** Males become sexually mature early in their second year, females in their second or third year. They mate in June and a single young is born in November. Newborn suckle from a teat under the mother's armpit and are carried in flight, clinging to a false pubic teat, legs wrapped around her chest, until well-furred at 4–5 weeks. They are then left clustered in the maternity cave while the mother forages, and are weaned at 8 weeks. **Diet** Moths, beetles, cockroaches, flies, wasps, other flying insects. **Habitat** Rainforests, open forests and woodlands. **Traces** Insect remains beneath hunting perches. Scats are small, friable, about 5 mm long, composed of insect fragments, and deposited at roosting sites. **Status** Low risk nationally, but vulnerable in Vic. where there are only 3 maternity caves. Threatened by human disturbance of roosts, destruction of roost sites and feeding habitats.

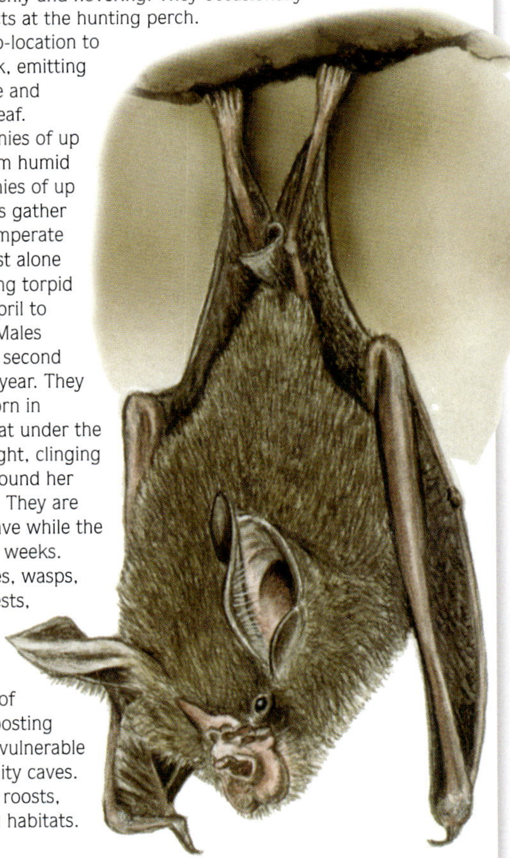

WEIGHT **7–14 g** HEAD–BODY **42–58 mm** FOREARM **44–52 mm** TAIL **22–28 mm**

Family **Hipposideridae** Species *Hipposideros ater*

Dusky Leaf-nosed Bat

The dusky leaf-nosed bat is a small, delicate, insectivorous bat with long soft fur, light grey above with blackish-brown to orange tips, and slightly paler below. It has small eyes, very large, broad and slightly pointed ears and a simple noseleaf with a horseshoe-shaped lower ridge above the snout. The tail protrudes slightly beyond the tail membrane.

Behaviour Active mainly at night, dusky leaf-nosed bats roost by day in caves, crevices, mines and sometimes tree-hollows, preferring small, hot, humid caves. They hang by their toes from the roof of the cave, 100–150 mm apart, wings against their sides and head hanging down. They roost alone or in colonies of 10–30 bats (although 300 have been recorded). In some areas they are preyed upon by ghost bats living in the same roost sites. They leave the roost at dusk and hawk for insects while flying slowly through the lower levels of the forest with a fluttering flight, changing direction frequently and hovering. They hunt close to their roost site in dense vegetation and gaps in the forest, sometimes in small groups, returning often to the cave entrance to rest and digest their prey. They have good sight, but use echo-location to navigate in darkness and detect flying insects, emitting ultrasonic signals through the nose and directing them with the noseleaf.

Development Dusky leaf-nosed bats mate from April to May and give birth to a single young from October to December. Newborn cling to one of 2 false teats in the mother's pubic region while she forages for the first 2 weeks, and suckle from a teat under each armpit. They then often cling to her shoulders and practise wing-flapping and echo-locating. They reach adult size after 6–8 weeks and accompany their mother on foraging expeditions. **Diet** Moths, mosquitoes and other flying insects caught on the wing. **Habitat** Rainforests, open forests, dry vine thickets, savannah woodlands and spinifex grasslands; from sea level to 1200 m. **Status** Low risk, but very sparse. Threatened by human disturbance of roosts and mining.

Weight **3–6 g** Head–body **33–50 mm** Forearm **34–41 mm** Tail **22–28 mm**

FAMILY **HIPPOSIDERIDAE** SPECIES *Hipposideros diadema*

DIADEM LEAF-NOSED BAT

The diadem leaf-nosed bat is a relatively large insectivorous bat with a fearsome set of teeth capable of inflicting a painful bite. It is easily identified by its pig-like face and distinctive white patches on the shoulders, contrasting with its reddish-brown to greyish-brown fur. Females tend to be more orange than males. The large ears are sharply-pointed, the eyes small, and the well-developed pink noseleaf has 4 deep pits in the upper section. The tail is usually enclosed in the tail membrane. The NT subspecies H.d. inornatus may be a separate species. **Behaviour** They shelter by day in small groups hanging by their toes from the ceilings of caves, old mines, sheds and road culverts, often with other species, particularly the eastern horseshoe bat. They also roost in large hollow trees and sometimes hang from branches. They emerge at dusk to hunt for a few hours in specific sites up to 10 km from the roost. They fly between hunting perches with limited manoeuvrability, following tracks in rainforests, alighting frequently on a branch to devour prey. In eucalypt forests they often hang upside down from a branch 1–3 m high, usually overlooking a creek, twisting from side-to-side with the ears in constant motion scanning ultrasonically for insects. If a swarm of insects comes close they catch a number in flight and store them in cheek pouches to be devoured later. Females congregate in caves forming maternity groups towards the end of the year. **Development** Sexually mature at 12 months, they breed in August or September and have a lifespan of up to 7 years in the wild and 12 years in captivity. Females give birth to a single large young weighing up to 13 g from November to January. It suckles from a teat in the mother's armpit and is carried with her on foraging trips until it is able to fly and feed alone at 2–3 months old. **Diet** Beetles, moths, leaf-hoppers, cockroaches and other large insects. **Habitat** Tropical rainforests and sclerophyll forests. **Status** Rare in Qld, vulnerable in the NT. Threatened by habitat destruction and disturbance by humans visiting roosting caves.

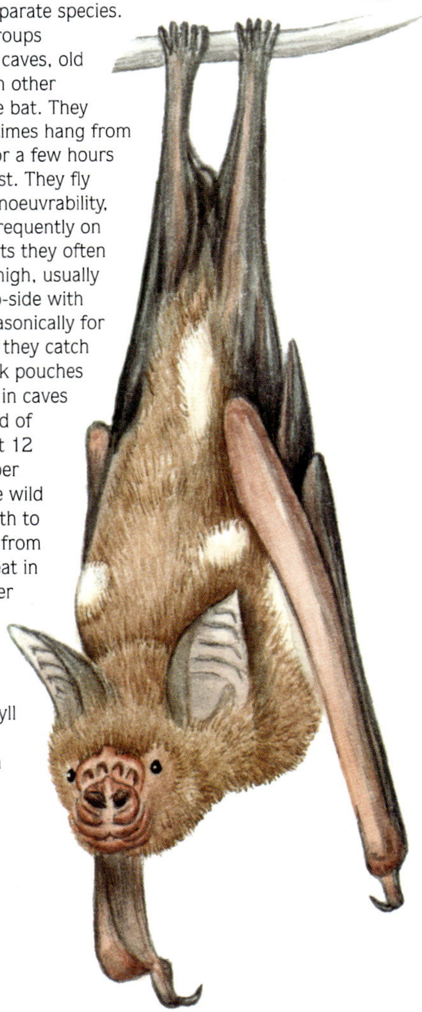

WEIGHT **34–53 g** HEAD–BODY **75–85 mm** FOREARM **68–85 mm** TAIL **30–40 mm**

FAMILY **HIPPOSIDERIDAE** SPECIES *Rhinonicteris aurantius*

ORANGE LEAF-NOSED BAT

The orange leaf-nosed bat is a small, insectivorous bat with fine golden fur, sometimes darker-tipped, with dark bands across the eyes. There are separate colonies of brown, yellow or all-white bats. It has small eyes, sharply pointed ears and a complex noseleaf above the snout with a horseshoe-shaped indented lower part, a scalloped upper section with deep pits and a forward projection in the centre. The tail protrudes slightly beyond the tail membrane. **Behaviour** Active mainly at night, orange leaf-nosed bats roost during the dry season in deep, very warm and humid caves in limestone formations. These caves usually have difficult access to deter predators such as the ghost bat. They roost in colonies ranging from 20 to 25,000 individuals, and disperse during the wet season, from November to February, probably to live in the forest. Despite their inability to maintain a high body temperature, they do not cluster together but space themselves 100–150 mm apart, suspended from the roof with the head hanging down or clinging to a wall with the forearms spread well apart, their wings folded against the sides of the body. At dusk they leave the roost to hunt for insects in nearby forests and woodlands, returning frequently to groom and digest their prey. They fly with fast wing beats, zig-zag close to the ground searching for insects, and are often seen following roads at night. They have good sight, but also use echo-location to detect flying insects up to 2 m away, emitting ultrasonic signals through the nose and directing them with the noseleaf. **Development** Females are sexually mature at 7 months, males at 18 months. They mate in July and females give birth to a single young in late December or early January. The young suckles from a teat in the mother's armpit, grows quickly and is weaned in late February. **Diet** Moths, beetles and flying termites, with some wasps, ants and other insects caught on the wing or gleaned from the foliage. **Habitat** Mangroves, spinifex grasslands, vine thickets, dense palm forests and woodlands. **Status** Vulnerable. Threatened by human disturbance of roosts and habitat destruction.

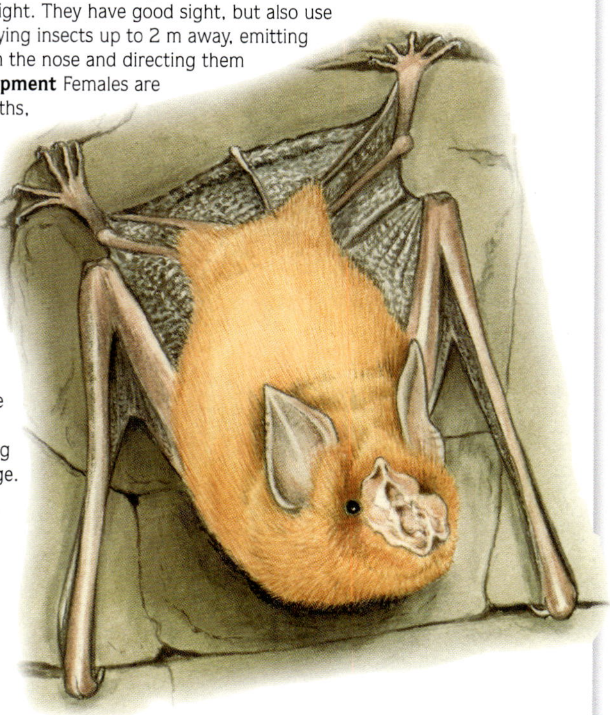

WEIGHT **6–12 g** HEAD–BODY **44–56 mm** FOREARM **45–51 mm** TAIL **23–29 mm**

FAMILY **EMBALLONURIDAE** SPECIES *Saccolaimus flaviventris*

YELLOW-BELLIED SHEATHTAIL-BAT

The yellow-bellied sheathtail-bat is the largest Australian sheathtail-bat. It has glossy black fur above and contrasting white or creamy-yellow belly fur. Both sexes have a pouched neck gland which is well-developed in males and reduced to a rudimentary fold of skin in females. The wings are long and narrow and the tips fold back over the membrane at rest. The tip of the tail is covered with bristles and appears to protrude through the tail-membrane, although it is connected. The tail slides freely, making it easier for the bat to use its hindlimbs. The head is dog-like, mostly naked, with black skin. It has thick, leathery, ribbed ears with a prominent lobe partially covering the ear aperture. Only the thumbs of the forelimbs have a claw. **Behaviour** Active at night, yellow-bellied sheathtail-bats are good climbers and fly fast and direct with slow wing beats, hunting high above the forest canopy and in forest clearings, often chasing each other in flight, possibly defending an aerial territory. They have good vision and are also able to navigate and detect flying insects at night by echo-location, emitting audible calls through the mouth while flying. They roost alone or in groups of up to 10 bats in buildings and large tree hollows, occasionally using the abandoned nests of sugar gliders. In treeless areas they have been found in mammal burrows. Secretions from the neck gland may be used by males to mark territories. In southern areas they migrate to warmer locations in winter. **Development** Births have been recorded from early December to mid-March. Females give birth to a single young that suckles from a teat in the mother's armpit. **Diet** Flying insects, particularly beetles, grasshoppers and moths. **Habitat** Most habitats from rainforests, sclerophyll forests and woodlands to grasslands and deserts. **Status** Low risk nationally, but vulnerable in NSW and rare in SA. Threatened by the logging of old trees.

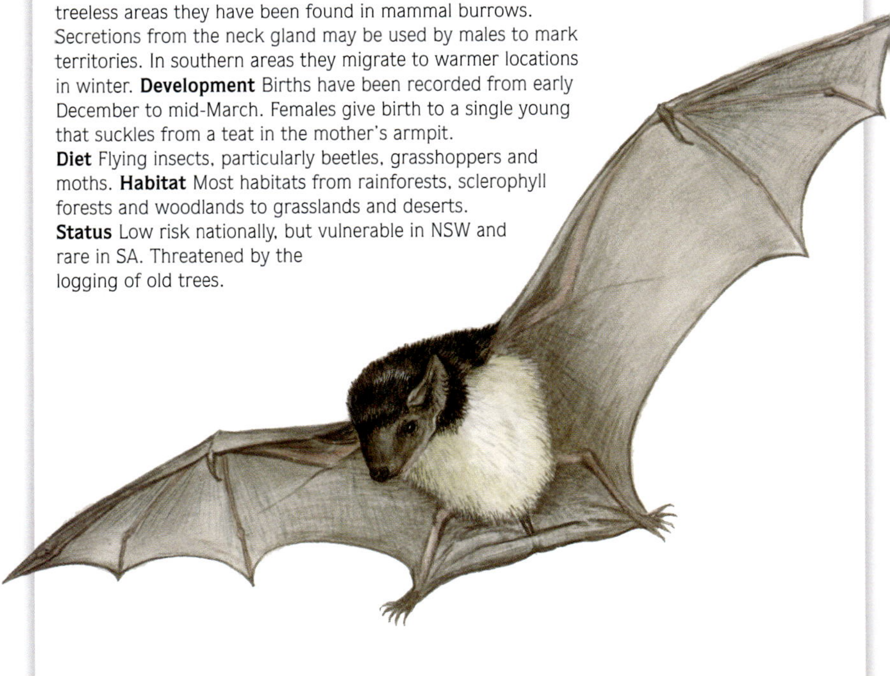

WEIGHT **30–60 g** HEAD–BODY **70–92 mm** FOREARM **65–82 mm** TAIL **20–35 mm**

FAMILY **EMBALLONURIDAE** SPECIES *Taphozous georgianus*

COMMON SHEATHTAIL-BAT

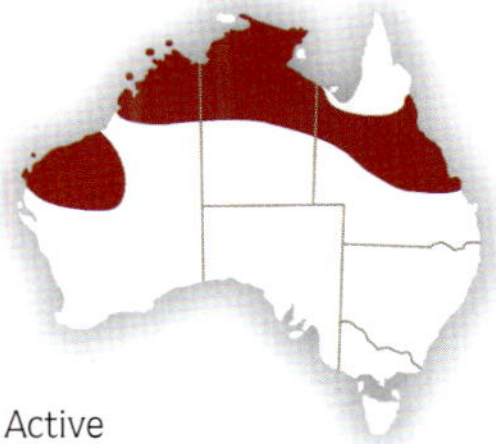

A widespread insectivorous bat, the common sheathtail-bat has dark brown fur above with a creamy-white base. It is light brown below with flecks of grey under the arms and sparse yellow hairs under the base of the tail. The tail protrudes through the tail-membrane and slides freely giving the hind limbs more freedom of movement and allowing the bat to move quadrupedally. The face is dog-like with large ridged ears and a prominent lobe partially covering the ear aperture. **Behaviour** Active at night, common sheathtail-bats roost alone or in small groups in caves, rock fissures and mine shafts. They roost in the twilight zone near the cave entrance, usually more than 500 mm apart, clinging to the walls fairly high above the ground. Individuals use many different caves, relocating regularly, and females may form maternity colonies. They have good vision and can detect insects up to 10 m away in complete darkness using echo-location, emitting high-pitched sounds through the mouth. They emerge at dusk and fly high and fast, often hunting in a grid pattern over shrubs and water, catching and eating insects in flight. They often rest on rocks while feeding and scuttle away to hide in a crevice if disturbed. Fat reserves are built up in summer and used in winter. On cold days they cluster together and become torpid to save energy. **Development** Females first mate at 9 months, males at 21 months. Only about 10 per cent of females and 3 per cent of males live more than 4 years. Males produce sperm in summer and autumn and store it until they mate from late August to September. A single young is born November–December. The newborn is well-developed, fully-furred with open eyes, and suckles from a teat in the mother's armpit. It is carried while the mother forages until able to fly at 3–4 weeks. Adult size is reached in 3–4 months. **Diet** Beetles and other insects. **Habitat** Wet and dry sclerophyll forests, paperbark forests, vine forests, monsoon forests, woodlands, spinifex and grasslands.
Status Low risk.

WEIGHT **16–51 g** HEAD-BODY **61–89 mm** FOREARM **61–75 mm** TAIL **20–40 mm**

FAMILY **EMBALLONURIDAE** SPECIES *Taphozous hilli*

HILL'S SHEATHTAIL-BAT

Often found roosting with the common sheathtail-bat, this small insectivorous bat was only recognised as a distinct species in 1980. It is a slightly different colour and the males have a small, pouched neck gland. The fur is rich clove-brown above and grades to light brown on the rump. The belly fur is light buff, tipped with olive-brown, and the wings are greyish-brown. The tail protrudes through the tail-membrane and slides freely giving the hind limbs more movement, allowing the bat to move quadrupedally. The face is dog-like and it has large ridged ears with a prominent lobe partially covering the ear aperture. **Behaviour** Active at night, Hill's sheathtail-bats roost in small colonies of up to 25 individuals in caves, rock fissures and mine shafts. They prefer to live near the cave entrance in the twilight zone, clinging to the walls with their clawed toes. When disturbed they scuttle around the walls like small crabs and hide in crevices. Pregnant females are reluctant to fly and roost in tiny crevices. They emerge from their roosts at dusk and hunt for insects above and around trees. They have good vision and can also detect insects at night using echo-location, emitting high-pitched sounds through the mouth. The pouch surrounding the male's neck gland enlarges in summer, and probably plays a role in mating behaviour. **Development** Males appear to be fertile all year. Females give birth to a single young between spring and late autumn. Newborn are large and well-furred, and suckle from a teat in the mother's armpit. **Diet** Insects. **Habitat** Arid and semi-arid woodlands; spinifex grasslands and acacia shrublands in rocky, hilly country. **Status** Low risk nationally, but rare in SA. Threatened by habitat destruction and disturbance of roosting caves.

WEIGHT **20–29 g** HEAD–BODY **63–81 mm** FOREARM **60–72 mm** TAIL **23–38 mm**

Family **Molossidae** | Species *Mormopterus beccarii*

Beccari's Freetail-bat

Often confused with the southern freetail-bat, this species is larger and occurs in more tropical parts of Australia. Beccari's freetail-bat is a robust, muscular, insectivorous bat with thick, elastic flight membranes and short, narrow, pointed wings. A large wrinkled upper lip fringed with stiff hairs and furrowed cheeks give it a bulldog-like appearance. It has a flattened head and body and hides in small cracks and crevices. The soft fur is dark grey-brown above and paler below. The large triangular ears are not joined above the head. The feet are stout with fleshy toes fringed with stiff curled hairs. The tail projects well beyond the tail-membrane. **Behaviour** Active mainly at night, Beccari's freetail-bats roost by day in colonies of around 50 bats mostly in tree hollows, but also in crevices, under loose bark and in buildings. They emerge at dusk and fly fast and direct with little manoeuvrability, hunting for insects above the forest canopy, along watercourses and over water. They use echo-location to detect insects in the dark, emitting high-pitched calls and homing in on flying insects up to 10 m away. They also have good sight and glean plant-eating bugs from the foliage, and scurry around on the ground and over tree trunks searching for flightless insects. **Development** Females give birth to a single young in the summer wet season. Pregnant females have been recorded from October to January and lactating females from November to January. The newborn suckles from a teat in the mother's armpit. **Diet** Flying and terrestrial insects, including beetles, moths, flies and grasshoppers. **Habitat** Wet and dry sclerophyll forests, woodlands, rainforests and urban areas. **Status** Low risk nationally, but vulnerable in NSW. Threatened by habitat destruction and loss of tree hollows.

Weight **10–19 g** | Head–body **54–65 mm** | Forearm **35–41 mm** | Tail **21–38 mm**

FAMILY **MOLOSSIDAE** SPECIES *Mormopterus loriae*

LITTLE NORTHERN FREETAIL-BAT

The little northern freetail-bat is a tiny, insectivorous bat with a very flat head and body, enabling it to hide in small cracks and crevices. It has soft, light brownish-grey fur with a whitish base, and is distinctly paler below with a yellowish throat and chin. It has large rounded ears which are not joined above the head and an upturned, hound-like snout with a protruding wrinkled upper lip that is fringed with stiff hairs. The feet are stout and have fleshy toes fringed with stiff curled hairs. The tail projects well beyond the tail-membrane. **Behaviour** Active mainly at night, little northern freetail-bats roost by day alone or in colonies of up to several hundred bats in tree hollows and crevices, under loose bark, and in caves and buildings. They emerge at dusk alone or in pairs to hunt for flying insects above and beside the forest canopy, over creeks and waterholes, and on the ground where they scurry around searching for terrestrial insects. They have short, narrow and pointed wings, enabling them to fly fast and direct but with limited manoeuvrability. They have good vision and are able to detect prey in the dark by echo-location using high-pitched calls. **Development** Females give birth to a single young in the summer wet season which suckles from a teat in the mother's armpit, until weaned around March. **Diet** Flying and terrestrial insects, including moths, beetles, cockroaches and ants. **Habitat** Rainforests, mangroves, monsoon forests, vine thickets, sclerophyll forests, paperbark forests and woodlands. **Status** Low risk.

WEIGHT **6–10 g** HEAD–BODY **42–55 mm** FOREARM **28–35 mm** TAIL **24–40 mm**

FAMILY **MOLOSSIDAE** SPECIES *Mormopterus norfolkensis*

EASTERN FREETAIL-BAT

This tiny bat was mistakenly assumed to come from Norfolk Island, and for many years was known as the Norfolk Island mastiff bat. It has an upturned snout and protruding wrinkled upper lip giving it a bulldog-like appearance. The soft fur is dark brownish-grey to reddish-brown above and slightly paler below. It has large triangular ears which are not joined above the head, and males have a glandular neck pouch. The feet have thick fleshy toes fringed with long, stiff, curled hairs, and the tail projects well beyond the tail-membrane. **Behaviour** Active mainly at night, eastern freetail-bats roost by day alone or in small colonies mostly in tree hollows, but also in crevices, under loose bark, and in caves and buildings. They cling to the surface with all four limbs, pressing close to the wall and bunching together when in groups. They hunt for insects above the forest canopy, in clearings and around forest edges, flying fast and direct but with limited manoeuvrability, travelling up to 12 km during the night. They have good sight, but are also able to detect flying insects in the dark using echo-location, emitting high-pitched calls. They also forage on the ground, scurrying around searching for terrestrial insects. **Development** Little is known about the development of eastern freetail bats. Breeding probably takes place in spring with a single young born in early summer. The newborn suckles from a teat in the mother's armpit.
Diet Flying and terrestrial insects.
Habitat Wet and dry sclerophyll forests, woodlands and occasionally rainforests.
Status Low risk in Qld, but vulnerable in NSW. Threatened by loss of habitat and lack of suitable tree hollows.

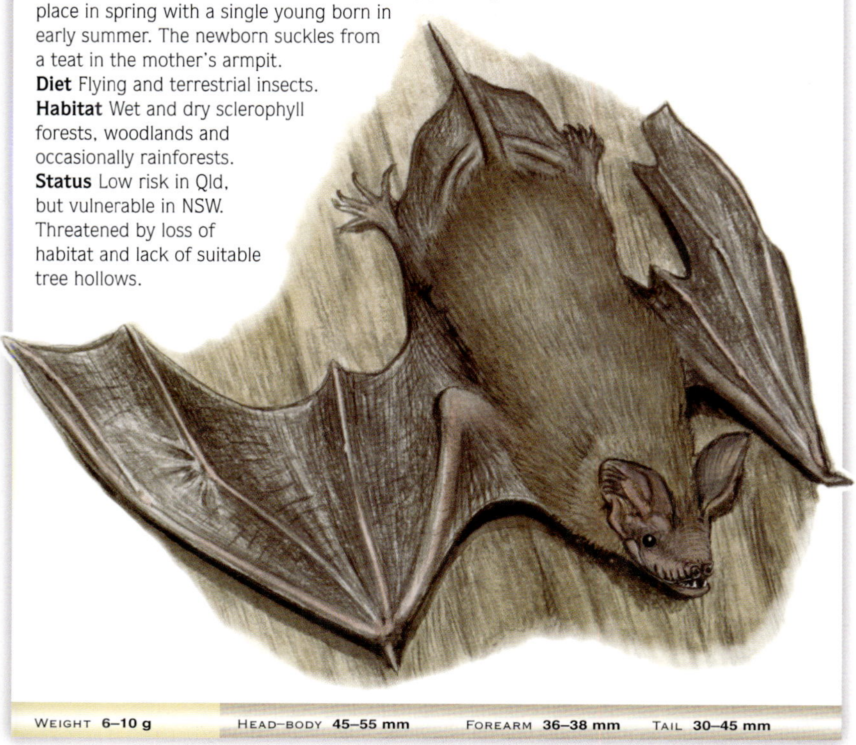

WEIGHT **6–10 g** HEAD–BODY **45–55 mm** FOREARM **36–38 mm** TAIL **30–45 mm**

FAMILY **MOLOSSIDAE** SPECIES *Mormopterus planiceps*

SOUTHERN FREETAIL-BAT

A small insectivorous bat with an upturned, hound-like snout, the southern freetail-bat has a flattened head and body enabling it to hide in small cracks and crevices. Its soft fur is grey to brownish-grey above and paler below. It has large triangular ears and a protruding, wrinkled upper lip with a dense fringe of stiff hairs. The feet are stout with fleshy toes fringed with stiff curled hairs, and the tail projects well beyond the tail-membrane. There are different forms of the southern freetail-bat. One has a penis less than 5 mm long, and is restricted to arid and semi arid areas of inland Australia. The other has shaggier, longer fur and a penis more than 9 mm long. **Behaviour** Active mainly at night, they fly fast and direct above the forest canopy, sometimes travelling more than 10 km to food sources. They have difficulty taking-off from the ground, and must climb to at least 1 m to launch themselves. They have good sight and can also detect insects in complete darkness using echo-location, emitting high-pitched audible calls. They forage in open areas, over creeks and waterholes, and on the ground, scurrying around searching for terrestrial insects, and can swim well if necessary. By day they roost alone or in colonies of 150 or more bats in tree hollows and crevices, and in buildings, often sharing roosts with other bat species. **Development** Females are sexually mature in their first year, males in their second year. Males produce mature sperm in late summer and autumn when food is plentiful, and usually mate in late autumn, although they can store viable sperm until the following spring, allowing them to mate throughout the winter. After mating, the female stores the sperm until August when she fertilises her ova. A single young is born from November to January and suckles from a teat in the mother's armpit. The young are independent at about 3 months. **Diet** Flying and terrestrial insects, including ants, beetles and flies. **Habitat** Dry sclerophyll forests, woodlands, mallee scrub, grasslands and deserts. **Status** Low risk.

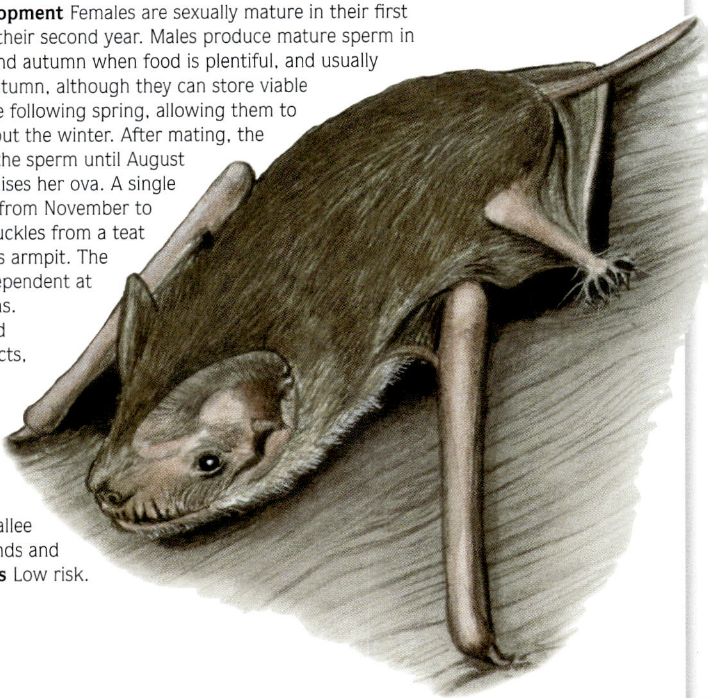

WEIGHT **7–15 g** HEAD–BODY **47–65 mm** FOREARM **30–40 mm** TAIL **25–40 mm**

FAMILY **MOLOSSIDAE** SPECIES *Tadarida australis*

WHITE-STRIPED FREETAIL-BAT

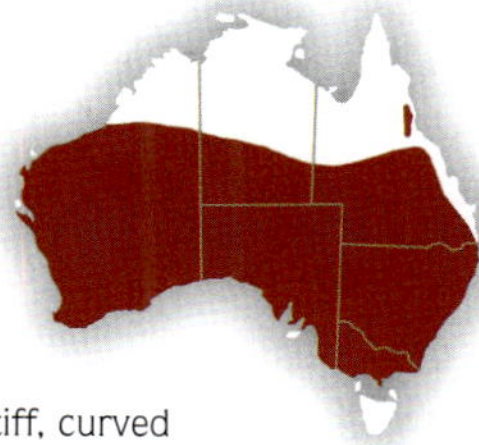

Although fierce-looking, the white-striped freetail-bat is quite docile when handled. The fur is short and dense, chocolate-brown to reddish-brown or black above, paler below with a distinct white stripe along the junction of the wings and body, and occasionally with white patches on the chest. It has fleshy, rounded, strongly-ribbed ears, almost touching above the head, and a deeply-wrinkled upper lip. Both sexes have a deep, wide throat pouch. The toes are thick, fleshy and fringed with long, stiff, curved hairs. The tail projects well beyond the tail-membrane and the wings are very compact when folded, allowing free movement of the limbs and agile movements on the ground. **Behaviour** Active at night, white-striped freetail-bats are fast, direct fliers with poor manoeuvrability, and tend to hunt in open areas and high above the canopy, sometimes covering large distances searching for food. They have difficulty taking off from the ground, and often crawl onto a rock or tree to launch themselves. They have good sight, and can also detect insects at night by echo-location, emitting regular, high-pitched, metallic, 'ting ting ting' calls. They also forage on the ground, scurrying around searching for terrestrial insects. They roost in tree hollows (rarely in rock caves or buildings), either alone or in colonies of up to 300 bats. They use a network of daytime roosts in an area as large as 200 sq km, switching every few days, with occasional day-time and frequent night-time visits to a central, communal roost. Maternity colonies of up to several hundred bats use the larger cavities of hollow tree trunks. Southern populations are thought to migrate to warmer climates in winter. **Development** Females are sexually mature in their first year, and most females reproduce every year. Mating occurs in late winter and a single young is born from mid-December to the end of January. The infant suckles from a teat in the mother's armpit, and is weaned by mid-February. **Diet** Insects, predominantly moths, with ants and beetles taken from the ground. **Habitat** Sclerophyll forests, woodlands, tropical savannas, semi-arid habitats, farmlands and urban areas; to at least 1400 m. **Status** Low risk.

WEIGHT **26–48 g** HEAD–BODY **85–100 mm** FOREARM **57–65 mm** TAIL **40–55 mm**

FAMILY **VESPERTILIONIDAE** SPECIES *Miniopterus australis*

LITTLE BENTWING BAT

The little bentwing bat is so-named because the tip of its third finger (which supports the tip of the wing membrane) is exceptionally long and bends under the wing at rest. The fur is long and thick, greyish-black to fawn-brown above and paler below. It moults early in the year, and the fur gradually becomes browner. The forehead is high and well-furred, and the muzzle is short. The tail is enclosed in the tail-membrane and folded under the body at rest. **Behaviour** Little bentwing bats are active at night and roost by day in caves, mine tunnels, tree hollows and buildings, hanging from the roof or walls and often forming mixed clusters with common bentwing bats, particularly in the south. Colonies of up to 100,000 or more congregate in early spring. Most adult males and juveniles disperse in summer, leaving maternity colonies of females and their young in warm nursery caves. The maternity colony disperses in winter and in southern areas the bats move into cool overwintering caves and become torpid to save energy. Northern populations remain active all year. They have good vision and also use echo-location to navigate and detect insects in the dark, flying quickly with great manoeuvrability, hunting for flying insects in densely wooded areas between the understorey and forest canopy. **Development** Mating takes place in July or August, but embryonic development is delayed until September and a single young is born in December. The newborn suckles from a teat in the mother's armpit until weaned in February. Very young bats are left in massed clusters in the maternity cave while the mother forages at night. They have a potential lifespan of more than 5 years. **Diet** A variety of insects including crane flies, ants, moths and wasps. **Habitat** Wet and dry sclerophyll forests, paperbark swamps and rainforests. **Status** Low risk in Qld, but vulnerable in NSW. Threatened by mining and disturbance of roosts.

WEIGHT **5–7 g** HEAD–BODY **40–48 mm** FOREARM **36–41 mm** TAIL **39–48 mm**

FAMILY **VESPERTILIONIDAE** SPECIES *Miniopterus schreibersii*

COMMON BENTWING BAT

One of the world's most widely distributed mammals, the common bentwing bat is so-named because the exceptionally long tip of its third finger bends under the wing at rest. The fur is chocolate-brown above and paler below. Patches of bright-rufous fur appear during the annual moult, and old bats are gingerish. The forehead is high and well-furred, the muzzle short. The tail is enclosed in the tail-membrane and folded under the body at rest. **Behaviour** Nocturnal, they fly fast and direct at up to 50 kph, making sudden dives to take small flying insects, generally hunting above the forest canopy, and travelling up to 70 km overnight. They have good vision and combine this with echo-location and hearing to detect insects in the dark. They roost in caves, mine tunnels and occasionally buildings, hanging from the roof or walls in separate massed clusters of the same sex or age group. Populations use the same roosts every year. Both sexes congregate in spring, and pregnant females join colonies of 100 to 150,000 bats in warm, humid, maternity caves where they raise their young. Colonies disperse in February and March, and individual bats migrate to winter roosts within about 300 km of maternity caves. In cold climates they become torpid for long periods to save energy. **Development** Both sexes reproduce in their second year. They mate in September in tropical regions and May–June in temperate regions (although the embryo remains dormant until August). A single, naked young is born in December–January (6 weeks earlier in South Australia), and suckles from a teat in the mother's armpit. Young are left in clusters on the cave roof while the mother forages. They can fly at 7 weeks, reach adult size at 10 weeks, and may live to 18 years or more. **Diet** Moths and other flying insects. **Habitat** Rainforests, wet and dry sclerophyll forests, paperbark forests, open woodlands and grasslands. **Status** Low risk nationally, but vulnerable in NSW. Threatened by disturbance and damage to roosting caves and predation by feral animals.

WEIGHT **9–20 g** HEAD–BODY **47–63 mm** FOREARM **43–51 mm** TAIL **50–60 mm**

FAMILY **VESPERTILIONIDAE** SPECIES *Nyctophilus arnhemensis*

ARNHEM LONG-EARED BAT

A small, tropical, insectivorous bat, the northern long-eared bat is olive-brown to light brown above and pale brown to greyish-brown below, with a dark base to the fur. The head is foreshortened with relatively large eyes; long, ribbed ears, each with a short, broadly-triangular lower lobe; and a poorly-developed noseleaf forming a low ridge behind the nostrils. The tail is enclosed in the tail-membrane. **Behaviour** Active mainly at night, northern long-eared bats roost by day in tree hollows, under the loose bark of paperbark trees, inside Pandanus leaf clumps, among tree foliage and in buildings. They emerge at dusk, their broad wings allowing them to fly slowly and flutteringly with high manoeuvrability as they forage in the forest understorey and over water for flying insects. They also glean flightless insects from the foliage and the ground. They have good vision and combine this with echo-location to navigate and detect insects in the dark.
Development Little is known about their development. Females give birth to 1–2 young in the dry season from October to February. Newborn suckle from teats in the mother's armpits.
Diet Flying and terrestrial insects.
Habitat Tropical rainforests, mangroves, woodlands and wet sclerophyll forests with an annual rainfall exceeding 500 mm.
Status Low risk.

WEIGHT **6–8 g** HEAD–BODY **40–59 mm** FOREARM **35–40 mm** TAIL **35–43 mm**

FAMILY **VESPERTILIONIDAE** SPECIES *Nyctophilus bifax*

NORTHERN LONG-EARED BAT

The northern long-eared bat is a small insectivorous bat with light brown or tawny-brown fur above and greyish-brown fur below. It has a foreshortened head with relatively large eyes and long ribbed ears, each with a short, broadly triangular lower lobe partially covering the ear aperture. A poorly-developed noseleaf forms a low ridge behind the nostrils. The tail is enclosed in the tail-membrane.

Behaviour Active mainly at night, northern long-eared bats roost by day in tree hollows, under peeling bark, among epiphytes and in dense foliage, emerging at dusk to hunt for insects in the forest understorey. Their broad wings allow them to fly slowly with high manoeuvrability and to hover and take off from the ground. They usually hang from trees 5–10 m high, rotate back and forth and make short, fluttering flights every few minutes to catch insects in flight, glean them from the foliage or from the ground. They have good vision and also use low-intensity echo-location calls to navigate and detect insects in the dark. Individuals change roosts frequently within a small area, often roost communally and move from summer roosts near the rainforest edge to warmer winter roosts deep in the forest. **Development** They mate in May and females usually give birth to twins in October. Newborn suckle from teats in the mother's armpits and are carried clinging to her fur while she forages for the first few days. They are then left at the maternity roost until weaned and able to hunt. Females often move their young to different roost sites. **Diet** Flying and terrestrial insects. **Habitat** Rainforests, paperbark forests, wet and dry sclerophyll forests and woodlands; often near watercourses. **Status** Low risk nationally, but vulnerable in NSW. Threatened by habitat destruction.

WEIGHT **6–14 g** HEAD–BODY **35–56 mm** FOREARM **32–46 mm** TAIL **34–46 mm**

FAMILY **VESPERTILIONIDAE** SPECIES *Nyctophilus geoffroyi*

LESSER LONG-EARED BAT

One of Australia's most widespread bats, the lesser long-eared bat is a tiny insectivorous bat with long fluffy fur, light grey-brown on the back and paler on the belly. It has small eyes, a small low noseleaf with a central Y-shaped groove projecting above the snout. The long, ribbed ears are joined above the head and have a short lower lobe. The tail is enclosed in the tail-membrane. **Behaviour** Active mainly at night, lesser long-eared bats leave their roost just after dark to hunt for insects, travelling 11 km or more overnight. They fly slowly with great manoeuvrability, foraging in the understorey, hovering over leaves, landing to take flightless insects and lifting off vertically from the ground. They have good vision and hearing and combine these senses with echo-location to detect prey in the dark. Tree hollows and crevices beneath loose bark are common roosts, but caves, abandoned bird nests, rocks and buildings are frequently used. They use a number of roost sites in a particular area, usually roosting alone or in groups of 2–3, although colonies of 200 or more have been found. Pregnant females form maternity colonies of about 10–15 bats in spring, sometimes with an adult male, and disperse in autumn when the young are independent. In cold southern areas they hibernate through winter or become torpid during the day to save energy. **Development** Males produce sperm in spring and summer and store it until mating in March–May. Females store the sperm through the winter, fertilising their ova in late August–September. They give birth, usually to twins, in October–November. Newborn suckle from teats in the mother's armpits and cling to her until well-furred at about 4 weeks. They are then left at the roost while the mother forages, and are weaned at 6–8 weeks. **Diet** Flying and terrestrial insects, particularly moths. **Habitat** Most habitats except tropical and subtropical rainforests, from sea level to 1600 m. **Status** Low risk.

WEIGHT **6–12 g** HEAD–BODY **38–50 mm** FOREARM **30–41 mm** TAIL **30–50 mm**

Family **Vespertilionidae** | Species *Nyctophilus gouldi*

Gould's Long-eared Bat

Gould's long-eared bat used to be considered a subspecies of the greater long-eared bat, which it closely resembles, although it is smaller. Their ranges overlap, but Gould's long-eared bat prefers wetter, more forested habitats. It is slate-grey to grey-brown above and ash-grey below, mottled with light buff. It has small eyes, long ribbed ears joined above the head with a short, broad, triangular lower lobe partially covering the ear aperture. A poorly-developed noseleaf forms a ridge behind the nostrils. The tail is enclosed in the tail-membrane. **Behaviour** Active mainly after dusk and in the early morning, their broad wings allow them to fly slowly with great manoeuvrability. They forage in the understorey, taking insects on the wing and gleaning insects from the foliage and the ground, catching prey in their wings, tail membrane or mouth. They have good vision and combine this with hearing and echo-location to navigate and detect insects, usually staying within 2 km of the roost. They roost alone or in colonies of up to 20 in tree hollows and under loose bark, preferring large trees, often changing roosts daily within a group of trees. In southern areas they spend the colder months in torpor, arousing to forage on warmer nights. Females form maternity colonies. Before leaving the roost, lactating mothers mark their young by smearing them with a yellow liquid secreted from glands on the muzzle, and rub faces with them on return. Females also use social calls to identify and maintain close bonding with their young for the first 3–4 months. **Development** Females are sexually active at 7–9 months, males at 12–15 months. Males produce sperm in summer and store it until mating in April–June. Females store the sperm and fertilise their ova in September–October, giving birth to 1–2 young in October–December. Newborn suckle from teats in the mother's armpits and cling to her for the first 3–5 days while she forages at night, and are then left in the roost. They are born naked, are fully furred by 10 days, begin to fly at 4 weeks and are weaned at 6 weeks.

Diet Moths, other insects and spiders.

Habitat Tall wet (and occasionally dry) sclerophyll forests, woodlands and subtropical rainforests; to at least 1250 m.

Status Low risk nationally, but endangered in SA. Threatened by habitat destruction.

Weight **5–17 g** | Head-body **50–70 mm** | Forearm **34–48 mm** | Tail **40–60 mm**

FAMILY **VESPERTILIONIDAE** SPECIES *Chalinolobus dwyeri*

LARGE-EARED PIED BAT

The large-eared pied bat is an insectivorous bat with long prominent ears and glossy black fur with a white V-shaped stripe in the pubic region beneath the wings and tail membrane. It has small eyes, a high forehead and a short muzzle with glandular swellings. The ears have a broad rounded lobe covering the base of the aperture. There is a horizontal lobe or wattle between the ear and corner of the mouth. The tail is enclosed in the tail-membrane. **Behaviour** Active after sunset and through the night, large-eared pied bats roost by day on the ceilings of caves in the twilight zone near the cave entrance, also in crevices in cliffs, mine tunnels and sometimes in the abandoned, bottle-shaped mud nests of fairy martins. Mixed colonies of 20–40 bats congregate in maternity roosts in early spring, huddling together for warmth, and returning to the same roost every year. Most males leave before the young are born and the roost is abandoned through late autumn and winter. They become torpid or hibernate during the coldest months to conserve energy when food is scarce. They have short, broad wings, fly fairly slowly with moderate manoeuvrability and hunt for small flying insects below the forest canopy. They have good sight and use echo-location to navigate and detect insects in the dark. Males secrete a milky substance from glands on either side of the muzzle during the breeding season. **Development** Females become sexually mature at about 12 months and give birth to 1–2 young in late November or early December. Newborn suckle from teats in the mother's armpits, are weaned in late January and are independent by late February. Juveniles leave the nursery colony by March. **Diet** Flying insects. **Habitat** Found in areas with extensive cliffs and caves close to dry sclerophyll forests and woodlands; to about 1500 m. **Status** Vulnerable. Threatened by logging, land clearing and disturbance of cave roosts.

WEIGHT **7–10 g** HEAD–BODY **45–55 mm** FOREARM **38–42 mm** TAIL **40–47 mm**

Family **Vespertilionidae** | Species *Chalinolobus gouldii*

Gould's Wattled Bat

Able to exploit a wide range of roosting sites, Gould's wattled bat is found in most of Australia. It has soft, dense fur, dark chocolate-brown to black above, often grading to brown towards the rump, and slightly lighter below. The muzzle is short and the forehead high. The ears are short and broad with a prominent rounded lobe partially covering the ear aperture. A loose flap of skin or wattle hangs down at the corner of the mouth, and the tail is enclosed in the tail-membrane. **Behaviour** Active from dusk to dawn, Gould's wattled bats take insects in flight or from the foliage, utilising open spaces above and below the canopy. They forage up to 17 km from their roost site, flying fast and direct, with sudden zig-zag changes of course, on long narrow wings. Roosts are in tree hollows, under loose bark, in rock crevices and buildings. Males are generally solitary, but females usually form colonies of 8–40 bats (200 have been recorded in a building), and individuals move frequently between roost sites within a particular area. Most juveniles disperse within 3 months of independence. In cold climates when food is scarce they become torpid to save energy, and in Tasmania they hibernate from late autumn to early spring. They have good sight and combine this with acute hearing and echo-location to detect insects in the dark, emitting an audible high-pitched sound in flight. Other vocalisations include buzzes, squeaks and chirps. **Development** Females are able to breed in their first year and mate at the beginning of winter, although the sperm is stored until they ovulate at the end of winter. Twins are born from September to October in the north and November to December in the south, and attach to teats in the mother's armpit. They grow rapidly, reaching adult size in one month, and are carried by the mother until well-furred, when they are left at the roost while she forages. The young are independent at 6 weeks and may live to 5 years or more. **Diet** A wide range of flying and crawling insects. **Habitat** Open and closed forests, mallee, tall shrubland, urban areas; from deserts to alpine regions to 1500 m. **Status** Low risk.

Weight **6–20 g** | Head–body **46–60 mm** | Forearm **35–50 mm** | Tail **31–50 mm**

FAMILY **VESPERTILIONIDAE** SPECIES *Chalinolobus morio*

CHOCOLATE WATTLED BAT

The chocolate wattled bat is an insectivorous bat with soft, dense, chocolate-brown fur and small lobes or wattles at the corner of its mouth. It has small eyes, a short muzzle and a high forehead with a ridge of fur between the eyes and snout. The ears are relatively small and rounded. The tail protrudes slightly beyond the edge of the tail-membrane. **Behaviour** Active between dusk and dawn, chocolate wattled bats fly fast between the tree canopy and understorey, sweeping low over forest trails hunting for insects. They have long narrow wings and fly fast and direct with good manoeuvrability, dropping suddenly on prey. They roost in tree hollows, abandoned bird nests, rock crevices, buildings and caves. Males usually roost alone, while females form colonies of up to several hundred bats, staying within a home range of up to 5 km. In cold climates they hibernate when food is short, entering hibernation later and emerging earlier than other small forest bats to take advantage of food resources. They have good sight and acute hearing, and use echo-location to detect insects in the dark. **Development** Males produce sperm in spring and summer and store it until mating in autumn. The female stores this sperm until ovulation and fertilisation in late winter, giving birth to a single young (occasionally twins) from mid-September to mid-November. Newborn suckle from teats in the mother's armpits, are carried until well-furred and become independent in January or February. **Diet** Flying insects, mainly moths and beetles. **Habitat** Wet and dry sclerophyll forests, rainforests, dry scrub and woodlands; to about 1600 m. **Status** Low risk.

WEIGHT **8–11 g**	HEAD–BODY **49–61 mm**	FOREARM **34–43 mm**	TAIL **39–50 mm**

Family **Vespertilionidae** Species *Chalinolobus nigrogriseus*

Hoary Wattled Bat

A small insectivorous bat, the hoary wattled bat takes its name from the white tips of its blackish-grey fur that give it a frosted or hoary appearance. It has a foreshortened head with small eyes, a high forehead, and relatively short rounded ears with a small lobe or wattle between the bottom of the ear and the mouth. The tail is enclosed in the tail-membrane. **Behaviour** Active mainly at night, hoary wattled bats roost by day in tree hollows and rock crevices. They are among the first bats to emerge from the roost after sunset, flying moderately fast and direct, hunting close to the ground in clearings, over small water bodies or above the tree canopy. They are very manoeuvrable, deviating suddenly from their flight path to catch evasive insects on the wing. Insects are gleaned from the foliage and they also land and crawl quickly after insects on the ground. They have good sight and can use echo-location in the dark to navigate and detect insects up to 5 m away. **Development** Little is known about their development. Females probably give birth to 1–2 young from October to November. Newborn suckle from teats in the mother's armpits and are weaned by January. **Diet** Moths, flies, beetles, crickets, spiders, ants and other invertebrates. **Habitat** Wet and dry sclerophyll forests, rainforests, woodlands and heaths. **Status** Low risk nationally, but vulnerable in NSW. Threatened by habitat destruction.

Weight **4–10 g** Head–body **39–48 mm** Forearm **30–40 mm** Tail **29–39 mm**

FAMILY **VESPERTILIONIDAE** SPECIES *Falsistrellus tasmaniensis*

EASTERN FALSE PIPISTRELLE

The eastern false pipistrelle is a relatively large insectivorous bat with long soft fur, rich-brown to reddish-brown above and slightly paler below. It has small eyes, a short, sparsely-haired muzzle with glandular swellings on the sides, long narrow ears with rounded tips, a notch on the outer edge and a narrow sharply-pointed lobe partially covering the aperture. It has a long tail that protrudes slightly beyond the tail-membrane. **Behaviour** Active mainly at night, eastern false pipistrelles fly fast and direct with limited manoeuvrability, foraging above or just below the tree canopy, up to 12 km from their roost. They attack insects on the wing by sudden changes in course, rather than by acrobatic movements. They have good sight, acute hearing and the ability to pin-point insects using echo-location. They generally roost in tree hollows but are sometimes found under loose bark, in caves and abandoned buildings, usually in single-sex colonies of 3–36 bats. In cooler mainland areas they migrate from the highlands to warmer coastal areas in winter, and in Tasmania they appear to hibernate from late autumn to early spring. **Development** Males produce sperm in the late summer or autumn when food supplies are plentiful, store it through the colder months and mate in late spring and early summer. A single young is born in December and suckles from a teat in the mother's armpit. **Diet** Moths, beetles, bugs, ants, flies and other insects. **Habitat** Wet forests and woodlands, preferring gullies and highland areas up to 1500 m. **Status** Low risk nationally, but vulnerable in NSW. Threatened by disturbance of roost sites and logging.

WEIGHT **14–29 g** HEAD–BODY **55–70 mm** FOREARM **45–56 mm** TAIL **40–51 mm**

Family **Vespertilionidae** Species *Myotis macropus*

Large-footed Myotis

A small bat with unusual foraging and social behaviour, the large-footed myotis is grey-brown to dark brown above, slightly paler on the belly and frosted with silver-grey on the chest. Older bats become ginger. Albinism and partial albinism are frequently seen. The ears are large and funnel-shaped with a long pointed lobe partially covering the aperture. The feet are large with long, widely-spaced, hairy toes with sharp curved claws and a very long ankle bone. The tail is enclosed in the tail-membrane. **Behaviour** Active at night, large-footed myotis bats hunt alone or in small groups, over creeks, rivers, estuaries and lakes, flying slowly with great manoeuvrability, spiralling downward to catch flying insects. They also rake aquatic insects and small fish from the water with their large clawed feet, or scoop them up with their tail membrane. Prey is detected using a combination of good eyesight, acute hearing and echo-location. They roost close to freshwater in caves, tunnels, buildings, beneath bridges, in tree hollows, among dense foliage and pandanus leaves. Colonies generally comprise 10–15 bats, but 200 have been recorded. In the breeding season adult males defend a tiny exclusive roosting site and control a harem of up to 12 females; older males have scarred ears from fighting. In southern areas they become torpid for periods during winter. **Development** Sexual activity begins in the first year of life. A single litter is produced in November–December in the south. In subtropical regions at least 2 litters are produced in early October and late January, while in the tropics 3 successive litters may be produced in one year. The single young suckles for about 8 weeks from a teat in the mother's armpit, and remains with her until independent 3–4 weeks later. **Diet** Insects and small fish. **Habitat** Sclerophyll forests, mangroves, paperbark swamps, woodlands and rainforests; near slow-flowing creeks, lakes and estuaries. **Status** Low risk nationally, but endangered in SA. Threatened by habitat destruction and poor water quality.

Weight **5–12 g** Head–body **35–60 mm** Forearm **36–43 mm** Tail **33–42 mm**

FAMILY **VESPERTILIONIDAE** SPECIES *Pipistrellus westralis*

NORTHERN PIPISTRELLE

A tiny bat with a delicate body, the northern pipistrelle is sometimes seen in urban areas in northern Australia catching flying insects around street lamps. It is dark brown to black above and lighter below. The head has a flattened crown and a sparsely-haired muzzle with glandular swellings on the sides. The ears are broadly triangular with rounded tips and have a triangular lobe partially covering the aperture. The upper jaw has two incisor teeth almost equal in length with slight indentations. The tail is enclosed in the tail-membrane. **Behaviour** Active mainly at night, northern pipistrelles roost by day in tree hollows and crevices, among dead palm fronds, below the leaf bases of pandans, in mines, caves and under house roofs. They land upright and turn over to hang by their feet or cling to vertical surfaces. They emerge after sunset to hunt for insects, following the outer foliage of the forest and flying below the canopy, above small creeks and tidal waterways in mangrove communities. They have a fast, fluttering flying pattern, swerving rapidly to take flying insects detected by sight or by echo-location in the dark. **Development** Northern pipistrelles breed year round. Females give birth to a single young that suckles from a teat in the mother's armpit. **Diet** Moths, beetles and other flying insects. **Habitat** Mangroves, pandan thickets, paperbark swamps and urban areas in coastal and near-coastal sites. **Status** Low risk.

WEIGHT **2.7–3.3 g** HEAD–BODY **34–43 mm** FOREARM **27–31 mm** TAIL **29–37 mm**

FAMILY **VESPERTILIONIDAE** SPECIES *Scoteanax rueppellii*

GREATER BROAD-NOSED BAT

This large, powerful bat has long, soft fur, dark reddish-brown to dark brown above and slightly paler below. Its broad head has a squarish outline when viewed from above, a low forehead, small eyes, broadly rounded ears with a narrow pointed lobe partially covering the ear aperture. The muzzle is sparsely-haired with glandular swellings, and only 2 upper incisor teeth are present. The tail is enclosed in the tail-membrane. **Behaviour** Active from dusk to dawn, greater broad-nosed bats are one of the first bats to emerge just after sunset. Their flight is slow and direct, and they hunt about 3–6 m above the ground, making only slight deviations from their flight path to catch moths, beetles and other large, slow-flying insects. They forage in forests and woodlands, utilising openings in the forest and corridors above creeks and small rivers. Hawking back and forth looking for prey, they also take small animals from the ground and foliage. They have good sight and use echo-location to navigate and detect prey in the dark. They roost by day in tree hollows and sometimes in the roof spaces of abandoned buildings. Pregnant females congregate at maternity sites in suitable tree hollows where they give birth and raise their young, apparently excluding males during the birth and raising of the young. **Development** Little is known about their development. A single young is born in January and suckles from a teat in the mother's armpit. **Diet** Large moths, beetles, bats, mice and other small vertebrates. **Habitat** Sclerophyll forests, rainforests and woodlands; preferring moist gullies below 500 m in the south and 800 m in the north. **Status** Low risk in Qld, but vulnerable in NSW. Threatened by habitat destruction.

WEIGHT **20–35 g** HEAD–BODY **63–95 mm** FOREARM **50–56 mm** TAIL **40–55 mm**

FAMILY **VESPERTILIONIDAE** SPECIES *Scotorepens balstoni*

INLAND BROAD-NOSED BAT

The inland broad-nosed bat is an insectivorous bat found over much of the arid and semi-arid regions of Australia. It is grey to grey-brown or chestnut-brown above and pale brown below. It has a broad head with a low forehead, a wide sparsely-haired muzzle with glandular swellings, slightly protruding nostrils and a V-shaped cleft in the lower lip. The eyes are small and the ears broadly rounded with a short, rounded lobe partially covering the aperture. The tail is enclosed in the tail-membrane and is arched under the body at rest and in flight. **Behaviour** Active soon after sunset, they fly relatively slowly with rapid diversions, only reacting to insects closer than 2 m. They forage below the canopy and at the edge of forests, skim repeatedly over water to catch small flying insects, and glean insects from the foliage and the ground. They have good sight and combine this with acute hearing and echo-location to detect prey. They roost in tree hollows, crevices and abandoned buildings, forming small colonies of up to 45 bats, often sharing roosts with larger colonies of southern freetail-bats. Pregnant females form maternity colonies. **Development** Mating takes place in April–May in the south and in September in the north. Females give birth to 1–2 young in October–November. Newborn are naked but well developed and able to walk. They suckle from teats in the mother's armpits, clinging to her for the first 10 days of life until they are too heavy to be carried. They are then left in the roost while she forages, and are independent at 5 weeks of age. **Diet** Insects including mosquitoes, termites, beetles, crickets, moths, flies and ants. **Habitat** Dry forests, woodlands, mallee scrub and farmlands; near permanent water. **Status** Low risk.

WEIGHT **6–14 g** HEAD–BODY **42–60 mm** FOREARM **30–41 mm** TAIL **29–42 mm**

FAMILY **VESPERTILIONIDAE** SPECIES *Scotorepens greyii*

LITTLE BROAD-NOSED BAT

Widely distributed in hot arid habitats, the little broad-nosed bat is an aggressive insectivorous bat with long soft fur, chestnut to grey-brown or ginger above, and paler below. It has a broad head with a low forehead and a squarish outline when viewed from above. It has small eyes, broadly rounded ears with a rounded lobe partially covering the aperture, a sparsely-haired muzzle with glandular swellings, and slightly protruding nostrils. The tail protrudes slightly from the tail-membrane.
Behaviour Active soon after sunset and at night, little broad-nosed bats roost by day in tree hollows, hollow fence posts and abandoned buildings, in pairs or colonies of up to 20 bats. Daily access to water is essential in the hot arid interior. They emerge at dusk and forage for flying and terrestrial insects in woodlands and plains mostly over waterholes and creeks. Prey is detected by a combination of good eyesight and echo-location. They fly quickly, darting frequently to catch insects, taking occasional drinks by skimming over the surface of still water.
Development Mating probably takes place in April, before the onset of winter in southern districts, with females giving birth in October and November, usually to twins. Mating may occur later in the tropics. Newborn suckle from teats in the mother's armpits and are able to fly by mid-December.
Diet Moths, flies, beetles, crickets, ants and termites. **Habitat** Wet and dry sclerophyll forests, monsoon forests, mangroves, paperbark forests, woodlands, arid areas, sand dunes and plains. **Status** Low risk.

WEIGHT **4–12 g** HEAD–BODY **37–55 mm** FOREARM **27–35 mm** TAIL **25–40 mm**

FAMILY **VESPERTILIONIDAE** SPECIES *Scotorepens sanborni*

NORTHERN BROAD-NOSED BAT

The northern broad-nosed bat is often found in roof spaces, and becomes aggressive if handled. It is a small, slender, insectivorous bat with long, soft, brown fur, tinged with red above and paler below. It has a broad head with a squarish outline when seen from above, a low forehead and small eyes. It has broadly rounded ears with a narrow pointed lobe partially covering the aperture, a sparsely-haired muzzle with glandular swellings and only two upper incisor teeth. The tail is enclosed in the tail-membrane. Females are usually larger than males. **Behaviour** Emerging soon after sunset, northern broad-nosed bats roost by day in tree hollows and occasionally in small spaces in buildings. In tree hollows they roost together in small colonies of 10–20 bats, while several hundred bats may congregate in the roof spaces of buildings. They are one of the first bats to appear at dusk, flying fast with much diving and darting to catch insects in flight. They usually fly below 5 m, hunting in the forest understorey, above watercourses, around calm coastal bays and urban street lights. They are often seen in summer skimming over still freshwater lakes and ponds, dipping their mouth in to drink. They have good sight and use echo-location to navigate and detect insects in the dark. **Development** Little is known about their development. They probably mate in May or June and give birth to a single young in late September or October which suckles from a teat in the mother's armpit. **Diet** Mosquitos, midges, mayflies and other small flying insects. **Habitat** Monsoon forests, rainforests, woodlands, mangroves and heaths. **Status** Low risk.

WEIGHT **5–9 g** HEAD–BODY **36–52 mm** FOREARM **27–36 mm** TAIL **27–39 mm**

FAMILY **VESPERTILIONIDAE** SPECIES *Vespadelus darlingtoni*

LARGE FOREST BAT

The large forest bat is a small, insectivorous bat. Its long dark brown to rusty-red fur has a brown base and light brown or grey tips, becoming very dark brown to black in older animals. It has dark brown to black wings, an abrupt forehead, rounded ears with a narrow lobe partially covering the aperture, and a membrane enclosing the tail.
Behaviour Active mainly at night, large forest bats fly fast and direct, making sudden gliding manoeuvres to attack flying insects. They hunt in open spaces below the forest canopy and around the top of the understorey, locating prey by echo-location, acute vision and hearing. They roost in tree hollows (preferring older smooth-barked trees), under loose bark, in crevices and abandoned buildings; alone, in groups of 5–6 or in colonies of up to 80 bats. Pregnant females form maternity colonies, often cohabiting with other bat species. They roost and forage within a home range that varies from less than 10 ha to more than 300 ha. In colder climates they hibernate from late autumn to early spring, emerging to feed on warm days when insects are abroad. **Development** Females are sexually mature in their first year, males in their second year. They mate in autumn. Females store the sperm until early spring when ovulation and fertilisation occur. A single young is born in November–December and suckles from a teat in the mother's armpit. The mother carries the young while she forages. When it becomes too heavy it is left at the roost. Young become independent in January–February. **Diet** Moths and other flying insects. **Habitat** Wet and dry sclerophyll forests and woodlands, rainforests, alpine moors; often near lakes; to about 1300 m. **Status** Low risk.

WEIGHT **6–10 g** HEAD–BODY **38–50 mm** FOREARM **31–37 mm** TAIL **29–40 mm**

FAMILY **VESPERTILIONIDAE** SPECIES *Vespadelus vulturnus*

LITTLE FOREST BAT

The little forest bat is a small insectivorous bat with dark to mid-grey fur on its back, sometimes tinged with brown, and a grey belly sometimes flecked with white. The skin on the upper forearms is lighter grey than on the wing membranes. It has a high forehead, a small furred muzzle and rounded ears with a narrow translucent white lobe partially covering the aperture. The tail is enclosed in the tail-membrane. **Behaviour** Active mainly at night, little forest bats hunt for flying insects below the forest canopy in the upper levels of the understorey, alighting at a perch to eat larger insects. Their flight is fast and fluttering with rapid changes of direction and spiralling dives to catch insects. Prey is located using echo-location, acute hearing and eyesight, and is sometimes caught in the wing or tail membrane. They roost in tree hollows and abandoned buildings, alone or in colonies of up to 120 or more bats, hanging from the walls or the backs of other bats. They forage and roost in areas averaging some 5 ha. Females form separate maternity colonies in early summer and mate with visiting males soon after giving birth. They disperse to mixed-sex winter roosts in late autumn, and become torpid for long periods in winter. **Development** Females are sexually mature in their first year, males in their second year. Mating takes place from March to the end of winter. Females store sperm until spring when ovulation and fertilisation occur. A single young is born in November or December and suckles for about 6 weeks from a teat in the mother's armpit. It is carried while she forages until it becomes too heavy, when it is then left at the roost until becoming independent at about 9 weeks of age. **Diet** A wide range of flying insects. **Habitat** Wet and dry sclerophyll forests and woodlands; mainly below 400 m, with patchy distribution to 1100 m. **Status** Low risk.

WEIGHT **4–6 g**	HEAD–BODY **34–50 mm**	FOREARM **24–33 mm**	TAIL **27–35 mm**

FAMILY **MURIDAE** SPECIES *Conilurus penicillatus*

BRUSH-TAILED TREE-RAT

Also known as the brush-tailed rabbit-rat, this attractive rodent of tropical northern Australia and southern New Guinea is solidly built and has a rabbit-like appearance with very long ears, large bulging eyes, a broad blunt head and elongated hind feet. It has long coarse fur, light grizzled grey-brown above and creamy-white below. The tail is grey-brown at the base with a black or sometimes white brushy tip, and overlapping scales. It has one pair of upper and lower gnawing incisor teeth and 3 pairs of molars on each side of the jaw. **Behaviour** Active at night, brush-tailed tree-rats sleep by day in tree hollows, crevices, in the crowns of pandanus trees or sand palms and buildings. They nest in family groups comprising a male and female and her young. They are agile climbers and spend most of their time foraging on the ground where they are very agile and bound rapidly along with the tail held up, flicking from side-to-side. They also search for food in trees, climbing up to 15 m above the ground. When disturbed they utter a threatening growl and climb the nearest tree. Other vocalisations include soft, high-pitched sounds.

Development Brush-tailed tree-rats have an extended breeding season lasting from March to October, timed to take advantage of food abundance during the late wet and mid dry seasons. Several litters are produced each season. Females have 4 teats and usually rear 2–3 young, born about 36 days after mating if she is not lactating, or 12 days later if she is lactating. Newborn are large and develop rapidly. They are fully-furred at 7 days, have their eyes open at 12 days and are weaned at about 20 days. Females can reproduce at only 6 weeks of age, although the average age is 11 weeks. They have a potential life span in captivity of 5 years. **Diet** Seeds, leaves, stems, fruit and termites. **Habitat** Tall, open eucalypt forests and woodlands, rainforests, pandanus scrub; often on the seashore near large casuarina trees. **Traces** Scats are small oval pellets about 2 mm across and 5 mm long, light to dark brown, containing fine plant fragments.

Status Low risk nationally, but vulnerable in the NT. Threatened by predation by feral cats and land clearing.

WEIGHT **102–242 g** HEAD–BODY **135–227 mm** TAIL **102–235 mm**

FAMILY **MURIDAE** SPECIES *Leggadina forresti*

FORREST'S MOUSE

Distinguished by its short stumpy tail, plump body, and blunt muzzle with many long whiskers, Forrest's mouse is a small placental mammal of the arid regions of Australia. The fur is thick, coarse and shiny, yellow-brown to grey above with scattered dark hairs, and pure white on the belly and feet. The tail is grey above and white below with overlapping scales. The pinkish-grey ears are short and rounded, the eyes large and bulging. It has one pair of upper and lower gnawing incisor teeth and 3 pairs of molars on each side of the jaw. **Behaviour** Active at night, Forrest's mice sleep during the heat of the day in grass-lined nests in shallow burrows 150–800 mm deep, or in cracks in the soil. Burrows may be up to 2.5 m long with several blind tunnels and a single nest chamber. Forrest's mice have no fixed home range and often travel more than 1 km overnight looking for food. Adults live alone or with their young, and females with young are intolerant of other adults. They become torpid when food is scarce to conserve energy and water. **Development** Forrest's mice breed after rainfall, and may rear several litters in succession in favourable conditions. Females have 4 teats and normally give birth to 3–4 young, 30–40 days after mating. The young are well-furred by 10 days, have their eyes open at 20 days, and are weaned at 28 days. They have a lifespan of up to 2 years. **Diet** Seeds, green vegetation, beetles and spiders. They can probably survive without drinking. **Habitat** Arid and semi-arid areas including tussock grasslands, low shrublands, mulga woodlands, monsoon forests, sand plains and clay pans. **Status** Low risk nationally, but vulnerable in NSW. Threatened by grazing stock and predation by feral animals.

WEIGHT **15–30 g** HEAD-BODY **65–105 mm** TAIL **50–75 mm**

Family **Muridae** | Species *Mastacomys fuscus*

Broad-toothed Rat

A stout, thickly-furred rodent of the alps and subalpine areas, the broad-toothed rat is named for its unusually broad molar teeth, enabling it to feed on tough plant material. The long fur is light to dark brown, grading to grey below. The tail is short with small bristles and overlapping scales. The head is wide with rounded ears and the hind feet are relatively long and slender. **Behaviour** Although mainly nocturnal, broad-toothed rats are often active during the day. They sleep in large well-insulated nests of finely-shredded grass under logs or dense vegetation, often in button grass. Females have overlapping home ranges up to about 1600 sq m and are very aggressive to males when suckling their young. Males have larger home ranges, up to about 2700 sq m, overlapping those of several females. They nest alone in summer and autumn, and congregate with up to 4 others in communal nests beneath the snow at the beginning of winter. Feeding sites are accessed via tunnels beneath the undergrowth. They stay warm and dry beneath a blanket of snow in winter and forage even on the coldest days. **Development** Females are sexually mature at about 10 months and breed from September to February in Tasmania, December to March in the Snowy Mountains. They have 4 teats and produce up to 2 litters of 1–3 young per season, born about 35 days after mating. Newborn are well-furred, and are dragged behind the mother if she is disturbed, clinging to her teats. They are weaned at about 5 weeks. **Diet** Stems and leaves of grasses and sedges, supplemented with bark, seeds and fungi. **Habitat** Cool, wet, alpine and subalpine heaths and swamp complexes to 2200 m; coastal grasslands, heaths and woodlands with dense groundcover. **Traces** Scats are deposited along runways and at nest entrances. They are fibrous, 3–5 mm across and 10–15 mm long, greenish-yellow when fresh and yellow-brown when dry. **Status** Low risk nationally, but vulnerable in NSW. Threatened by land clearing and predation by feral animals.

Weight **97–200 g** | Head–body **140–195 mm** | Tail **100–135 mm**

FAMILY **MURIDAE** SPECIES *Mesembriomys gouldii*

BLACK-FOOTED TREE-RAT

The black-footed tree-rat is a rabbit-sized placental mammal and one of Australia's largest rodents. It has long, coarse and shaggy fur, grey-brown flecked with black above, and creamy-white below. The tail is black with a white brushy tip and overlapping scales. It has long rounded ears, bulging eyes, and broad hind feet with strong claws and well-developed sole pads. It has one pair of upper and lower gnawing incisor teeth and 3 pairs of molars on each side of the jaw. **Behaviour** Active only at night, black-footed tree-rats are solitary, sleeping by day in nests in tree hollows, among palm fronds, in crevices and buildings. Individuals use several nest sites in different trees, preferring hollows about 100 mm diameter and 6–8 m above the ground. They are aggressive and defend their nest sites, uttering threatening grumbling and growling sounds before fighting. They are fast runners and agile climbers and may travel 500 m or more between refuge trees and feeding sites, often feeding on the ground on fallen fruits, rapidly ascending a tree if disturbed. **Development** Black-footed tree-rats breed throughout the year with a peak in the late dry season (August–September), although few births occur in the wetter months of January–March. Females have 4 teats and usually rear 2 young born 43–44 days after mating. Newborn are dragged by the mother clinging to her teats if she leaves the nest until they are 10–11 days old by which time they are fully furred with their eyes open. They are weaned at about 28 days and are fully grown at 11–12 weeks. They have a lifespan of up to 8 years and females are sexually mature at about 10 weeks. **Diet** Nuts, fruits (including pandanus fruits), flowers, insects and molluscs. **Habitat** Open forests and tropical woodlands with a grassy or shrubby understorey. **Status** Low risk.

WEIGHT **580–900 g** HEAD–BODY **250–310 mm** TAIL **320–415 mm**

FAMILY **MURIDAE** SPECIES *Mesembriomys macrurus*

GOLDEN-BACKED TREE-RAT

A rabbit-sized placental mammal, the golden-backed tree-rat is an arboreal rodent of tropical northern Australia. It has coarse fur, golden-brown on the back, grading to grey on the sides and white below. The base of the long tail is grey, the rest is white with a brushy tip and overlapping scales. The ears are long and rounded, the eyes bulging and the hind feet broad with strong claws and well-developed sole pads. It has one pair of upper and lower gnawing incisor teeth and three pairs of molars on each side of the jaw. **Behaviour** Active mainly at night, golden-backed tree-rats sleep by day in tree hollows and among dense foliage under spiky pandanus crowns in nests of loosely woven leaves. Although predominantly arboreal, they also forage on the ground and have been seen foraging just after sunrise along the high-tide line of an open beach. They can travel long distances rapidly, scampering through the grass and undergrowth with the tail held high. Pairs of adults and their immature offspring occupy a home range up to 600 m long. **Development** Females are sexually mature at 10 months and they probably breed year round. They have 4 teats and give birth to 1–3 (usually 2) young about 47 days after mating. The young are weaned at 6–7 weeks and are fully grown at 4 months. They have a lifespan of 7 years. **Diet** Flowers, fruits, seeds, grass, leaves and invertebrates. **Habitat** Open forests and woodlands with a grassy or shrubby understorey; mangroves, palm forests, pandanus clumps and vine thickets; in rugged country with more than 600 mm of rain annually. **Status** Vulnerable. Threatened by land clearing and pastoral activities.

WEIGHT **207–330 g** HEAD–BODY **180–245 mm** TAIL **290–360 mm**

Family **Muridae** | Species *Notomys alexis*

Spinifex Hopping-mouse

With its long hind legs, bipedal hopping gait and long tufted tail, the spinifex hopping-mouse is a small and distinctive placental mammal of the arid zone. The short fur is light brown to chestnut above and grey-white below. Both sexes have a small throat pouch of unknown function. All males and some females have a small, naked, glandular area on the chest. The ears are long, rounded and sparsely-haired. It has one pair of upper and lower gnawing incisor teeth and 3 pairs of molars on each side of the jaw. **Behaviour** Active at night, spinifex hopping-mice shelter from the heat of the day in deep, humid burrows. Burrows are up to 1 m below ground, and have several vertical shafts connected to a broad horizontal tunnel leading to a nest chamber lined with twigs, grass or shredded vegetable matter. They are highly social and 10 adults may share a burrow, plugging the entrance with sand to reduce evaporative water loss. Groups build adjacent burrow systems linked by runways, and cooperate in digging, rearing young and rejecting strangers. Secretions from the chest gland are probably used to scent-mark group members. They use a quadrupedal gait when moving slowly, but bound on their long hind feet with their brush-tipped tail trailing behind when moving fast. **Development** Sexually mature at about 11 weeks, they can reproduce at any time of year, with peaks in spring and after periods of good rainfall. Pairs lock during copulation and often struggle violently or turn back-to-back before separating. Females have 4 teats and rear litters of usually 3–4 young, born 32–35 days after mating. Young are left in the nest while the mother forages, and may be retrieved by any adult if they stray. They have their eyes open at about 20 days, are weaned at 4–5 weeks and live for up to 6 years in captivity. **Diet** Seeds, leaves, roots and insects. They can survive without drinking. **Habitat** Arid spinifex and tussock grasslands, woodlands, shrublands and desert sand dunes. **Traces** Scats are small brown pellets, about 2 mm across and 5 mm long, containing fine plant particles, and are scattered near burrows and runways. **Status** Low risk.

Weight **28–50 g** | Head–body **90–115 mm** | Tail **115–150 mm**

FAMILY **MURIDAE** SPECIES *Notomys cervinus*

FAWN HOPPING-MOUSE

Found in the arid gibber plains of central Australia, the fawn hopping-mouse has long hind legs, a distinctive bipedal hopping gait and a long, tufted tail. The fur is pale pinkish-fawn to grey above and white below. Males have a small, naked, raised, flat glandular area on the chest, 2–3 mm across, often stained by pale yellow flaky secretions in sexually active animals. The ears are long, rounded and sparsely-haired. It has one pair of upper and lower gnawing incisor teeth and 3 pairs of molars on each side of the jaw. **Behaviour** Active at night, fawn hopping-mice shelter during the day in simple, cool, humid burrows with 1–3 entrances. Burrows in gibber plains are up to 1 m deep and have up to 4 occupants. In hard claypans they are up to 300 mm deep and have up to 7 occupants. They live in family groups and forage alone within a few hundred metres of the burrow. Fawn hopping-mice are more aggressive than other hopping-mice, but avoid sandy areas which are the domain of the dusky hopping-mouse. The male's chest gland is probably used to scent-mark group members or areas within its home range. **Development** Sexually mature at about 6 months, they are opportunistic breeders, probably reproducing after periods of good rainfall, and undergoing periodic population increases. Females rear litters of usually 3 young born 38–43 days after mating, although pregnancy may be prolonged by up to 11 days if she is still suckling the previous litter. Newborn cling to the mother's teats and are dragged around by her if disturbed. They are well-furred at birth, have their eyes open at about 21 days and are weaned at 4 weeks. **Diet** Seeds, green plant matter and some insects. They can drink salty water and are able to survive without water. **Habitat** Gibber plains, tussock grasslands and low shrublands with hard clay and stony soils. **Status** Low risk in Qld, but endangered in SA and the NT. Threatened by grazing stock and predation by feral animals.

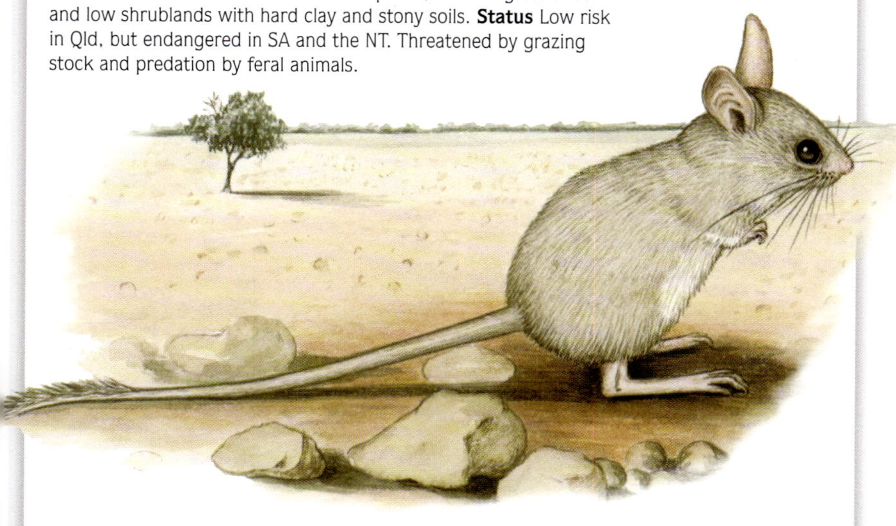

WEIGHT **30–50 g** HEAD–BODY **95–120 mm** TAIL **105–160 mm**

FAMILY **MURIDAE** SPECIES *Notomys mitchelli*

MITCHELL'S HOPPING-MOUSE

Distinguished by its long hind legs and bipedal hopping gait, Mitchell's hopping-mouse is fawn to dark grey above, often grizzled, and grey-white below. A wide tract of shiny white hairs runs from the throat to chest. The tail is brown or grey above and lighter below with a brush of dark hairs at the tip. The ears are long and oval-shaped with sparse brown hairs outside. Like other rodents it has one pair of upper and lower gnawing incisor teeth and 3 pairs of molars on each side of the jaw. **Behaviour** Active at night, Mitchell's hopping-mice shelter by day in a nest of grass or shredded vegetation in a deep burrow system (hollow logs are used in mallee woodlands). The nest chamber is at the end of a broad horizontal tunnel around 100–150 mm diameter and 1.5 m below ground, accessed by several vertical shafts (pop-holes) dug from below. Mixed groups of up to 8 adults build adjacent burrow systems up to 150 m apart. Each group uses several burrows and individuals may share a burrow to maintain high humidity and reduce water loss in desert conditions. Secretions from glands on the neck and chest are thought to be used to mark home ranges and members of the colony. They move with a bipedal hopping gait, which uses less energy than running. **Development** Sexually mature at about 3 months, they probably breed throughout the year in favourable conditions, rearing litters of 1–4 young. Pregnancy lasts 37–40 days, and the newborn cling tightly to the mother's 4 teats and are dragged behind her when she moves. They are well-furred at 7 days, open their eyes at 20 days, are weaned at 30–35 days, and live for up to 5 years in captivity. **Diet** Mostly seeds and fruit, supplemented by flowers, leaves, stems, roots and insects. **Habitat** Semi-arid mallee woodlands, dry heaths, dunefields and shrublands. **Traces** Scats are deposited near burrows and runways. They are small pellets, 1–2 mm across and about 5 mm long, containing fine plant fragments. **Status** Low risk.

WEIGHT **40–60 g** HEAD–BODY **100–130 mm** TAIL **140–160 mm**

FAMILY **MURIDAE** SPECIES *Pseudomys albocinereus*

ASH-GREY MOUSE

A small attractive rodent, the ash-grey mouse is restricted to the semi-arid regions of southwestern Australia. It has long soft fur, silver-grey tinged with fawn above, and white below. The nose, feet and tail are bright pink, the bulging eyes have prominent dark eyelashes, and the muzzle has long black or white whiskers. The ears are large and rounded, the tail slightly hairy with overlapping scales. It has one pair of upper and lower gnawing incisor teeth and 3 pairs of molars on each side of the jaw. **Behaviour** The ash-grey mouse is nocturnal, and forages on the ground at night for seeds and green vegetation, supplemented by insects in summer. During the day individuals or family groups, comprising a breeding pair and their growing young, sleep in a cool burrow up to 4 m long and about 600 mm below ground. Some burrows are complex with a number of interconnecting nesting chambers, vertical shafts, side tunnels and a second entrance. Several litters of various ages may be found in a single burrow. **Development** The ash-grey mouse breeds in spring in the western part of its range. In eastern districts breeding takes place when food is available, with peaks after rain. Sexual maturity is reached at about 10 months and they have a lifespan of 3 years. Females have 4 teats and usually rear 4 young born some 38 days after mating. The young open their eyes at 15 days and are weaned at about 40 days. Several litters may be raised in a season. **Diet** Seeds, berries, nectar, pollen, fungi, lichen and insects. It obtains sufficient water from its food. **Habitat** Semi-arid tussock grasslands, tall shrublands, open sandplains with heath vegetation and woodlands. **Traces** Piles of excavated sand at burrow entrances. Scats are small round to oval pellets, about 2 mm across, containing fine plant fragments. **Status** Low risk.

WEIGHT **21–40 g** HEAD–BODY **63–110 mm** TAIL **85–115 mm**

FAMILY **MURIDAE** SPECIES *Pseudomys australis*

PLAINS RAT

One of Australia's most attractive rodents, the plains rat is a desert species with a silver sheen to its soft thick fur. It is sandy-brown to grey above and white to cream below. It is quite bulky with large ears, bulging eyes and a well-furred tail, brown or grey above and white below with a lighter tip. It has one pair of upper and lower gnawing incisor teeth and 3 pairs of molars on each side of the jaw. **Behaviour** Active mainly at night, plains rats sleep by day in cool shallow burrows dug into the hard gibber, claypans, or in swampy or sandy ground. Burrows are about 1 m long and up to 300 mm deep, with a single nest chamber often stuffed with vegetation. Temporary burrows are also used for shelter. Up to 20 adults may occupy a single burrow outside the breeding season, although when breeding, a single male will share with 1–3 females. They gather in colonies occupying a group territory of about 1 ha with a number of burrows about 10 m apart, linked by a network of surface runways. They undergo large population fluctuations and colonies rapidly increase or decrease in size depending on food sources and predators. When alarmed they stand up on their hind legs and squeal. They have a fairly extensive vocal repertoire of chirps, screeches and whistles. **Development** Females are sexually mature at about 5 months, males at 6 months, with a lifespan of up to 8 years. They breed mostly in winter and spring, probably after heavy rain, sometimes rearing several litters in succession. Females have 4 teats and usually rear 3–4 young, born 30–31 days after mating. The young open their eyes at 15 days and are weaned at about 30 days. **Diet** Seeds supplemented with vegetable matter, and insects. They can survive without drinking. **Habitat** Arid gibber plains, ephemeral river flats, claypans, sand ridges with low shrubs. **Traces** Scats are found along runways and around burrow entrances. They are brown, cylindrical to oval, about 2 mm across and 3–8 mm long, containing insect fragments or fine powdery particles.
Status Endangered. Threatened by predation by feral animals and habitat disturbance by stock.

WEIGHT **50–80 g** HEAD–BODY **100–140 mm** TAIL **80–120 mm**

FAMILY **MURIDAE** SPECIES *Pseudomys delicatulus*

DELICATE MOUSE

A tiny tropical species, the delicate mouse is Australia's smallest native rodent, and is easily killed if handled. Dainty and graceful in appearance, it has yellow-brown to grey-brown fur above and is white or cream below. The nose and feet are bright pink, and the tail is slender with overlapping scales. The ears are quite large and rounded (10–13 mm long), the eyes large and bulging. It has one pair of upper and lower gnawing incisor teeth and 3 pairs of molars on each side of the jaw. **Behaviour** Active mostly in the hours before and after sunrise and sunset, the delicate mouse sleeps by day in a grass-lined nest in a burrow, in a termite mound, hollow log or under bark. In hard granite soils the burrows are short and shallow with false passages and a single nest chamber. In soft soil the burrows may be up to 400 mm deep with a single tunnel to 2 m long, terminating in a grass-lined spherical nest chamber. It uses different areas for day and night resting spots, often up to 200 m apart. **Development** The delicate mouse reaches sexual maturity at 10–11 months, and usually mates in June and July in the late wet season, producing several litters in favourable conditions. Females have 4 teats and normally give birth to 3–4 young, born 28–31 days after mating. Newborn are about 25 mm long and weigh only 1 g. Their eyes open at 20 days, and they are fully-furred and weaned at 30 days. **Diet** Seeds of native grasses, spinifex stalks. **Habitat** Coastal sand dunes with sparse vegetation of grasses, herbs and stunted trees; low woodlands, open scrublands and hummock grasslands. **Traces** Spoil heaps to 100 mm high around burrow entrances. **Status** Low risk nationally, but endangered in NSW. Threatened by feral animal predation.

WEIGHT **6–15 g** HEAD–BODY **55–75 mm** TAIL **55–80 mm**

FAMILY **MURIDAE** SPECIES *Pseudomys hermannsburgensis*

SANDY INLAND MOUSE

The sandy inland mouse is a slender, slightly-built rodent, very similar in appearance to the introduced house mouse, although it lacks the latter's distinctive musty smell, and has larger ears, eyes and tail. The fur is sandy-brown to grey above, grading to white below. It has one pair of upper and lower gnawing incisor teeth and 3 pairs of molars on each side of the jaw. **Behaviour** Active mainly at night, sandy inland mice sleep in a grass-lined nest in a cool deep burrow near the base of a shrub or tree. Burrows have a single tunnel up to 1 m long extending about 500 mm below the surface. They are gregarious and share burrows, although social tolerance depends on their breeding condition. Small groups of 4–6 breeding adults have been found in the same burrow, and 22 non-breeding adults in another burrow. Pregnant or lactating females become intolerant of other adults and have been recorded castrating or killing their mates. They forage mostly on the ground and will climb into shrubs and trees searching for food, climbing up to 1 m above the ground and travelling up to 14 km. They sometimes enter a state of torpor to conserve energy and water when food is scarce. They undergo large population fluctuations. **Development** Sexually mature at about 3 months, they breed after good rainfall, and produce several litters in favourable conditions. Females have 4 teats and normally give birth to 3–4 blind, naked young, 30–34 days after mating. At 20 days they are well-furred with their eyes open, and are independent at 30 days. **Diet** Seeds, roots, flowers, tubers, grass and invertebrates. They can survive without drinking. **Habitat** Woodlands, mallee shrublands, hummock grasslands, gibber plains and alluvial flats. **Traces** Scats and piles of soil at burrow entrances. Scats are tiny brown pellets about 1 mm across and 3 mm long, containing fine plant fragments. **Status** Low risk nationally but vulnerable in NSW. Threatened by feral animal predation and competition by grazing stock.

WEIGHT **8–15 g** HEAD–BODY **55–85 mm** TAIL **70–90 mm**

FAMILY **MURIDAE** SPECIES *Pseudomys higginsi*

LONG-TAILED MOUSE

Found only in Tasmania, the long-tailed mouse has very soft dense fur forming a prominent ridge between the eyes. It is dark grey-brown above fading to fawn on the sides and grey white below, and has a very long tail, dark grey above and white below, carried curved up well clear of the ground. The ears are large and round, the hind feet are long and slender and clothed with white hairs. It has one pair of upper and lower gnawing incisor teeth and 3 pairs of molars on each side of the jaw. **Behaviour** Active mainly at night in warmer weather, long-tailed mice often forage by day in winter, following runways and tunnels beneath the deep moss and forest debris. They sleep in nests of shredded bark or grass in a hole in a rotting stump or log, beneath leaf litter, or at the end of a short burrow. Placid, curious and highly sociable, they remain in a small home range and form permanent breeding pairs, sharing a nest with their offspring and defending a territory. Up to 10 families per hectare can be found in good habitats. They are good climbers, poor swimmers, leap randomly if threatened, and make a faint high-pitched whistle when disturbed. **Development** The breeding season extends from October to March, with 1–2 litters produced per year. Females have 4 teats and usually give birth to 3–4 young, born 31–32 days after mating. Newborn are well-furred and cling to their mother's teats or are left in the nest with the entrance temporarily plugged while the mother forages. If the mother is disturbed they cling to her teats and are dragged along behind her. They open their eyes at 15 days, are weaned at 25–30 days, leave the nest at about 3 months and may live to about 18 months. **Diet** Grass, seeds, fruit, lichen, ferns, insects and spiders. **Habitat** Rainforests, wet and dry sclerophyll forests and subalpine scree. **Traces** Scats are dark brown pellets about 3 mm across and 5 mm long, containing fine plant material. **Status** Low risk.

WEIGHT **50–90 g** HEAD–BODY **115–150 mm** TAIL **145–200 mm**

FAMILY **MURIDAE** SPECIES *Pseudomys nanus*

WESTERN CHESTNUT MOUSE

A small rodent of the tropical north of Australia, the western chestnut mouse has short limbs, a stout body and small, round ears. The fur is light fawn-orange above with many long dark brown hairs, light orange-brown on the flanks, and white on the belly. It has a pronounced light eye ring, and a naked tail with overlapping scales. Like other rodents it has one pair of upper and lower gnawing incisor teeth and 3 pairs of molars on each side of the jaw. **Behaviour** Despite its abundance, little information is available about the lifestyle of the western chestnut mouse. It is known to be active at night and to sleep by day in a grass nest. It makes repeated use of a number of regular nest sites within its home range and travels more than 400 m while foraging. In captivity they fight each other and make frequent high-pitched whistling calls when active. **Development** Western chestnut mice breed mostly towards the end of the wet season, but may breed at any time of year if conditions are favourable. Their habitat is subject to fires and flooding, and they probably rear several litters in rapid succession to take advantage of good conditions. Females have 4 teats and usually give birth to 3 young born 22–24 days after mating. The young are furred at 7 days, have their eyes open at 12 days and are weaned at about 21 days. **Diet** Native grasses and seeds. They may be able to survive without drinking, obtaining sufficient water from their food. **Habitat** Woodlands, tropical tussock grasslands and open forests with good ground cover on rocky or sandy soil. **Status** Low risk.

WEIGHT **25–50 g** HEAD–BODY **80–140 mm** TAIL **70–140 mm**

FAMILY **MURIDAE** SPECIES *Pseudomys novaehollandiae*

NEW HOLLAND MOUSE

Easily confused with the house mouse, but lacking its distinctive odour, this small placental mammal has larger ears and eyes. Its tail is dusky brown above and white below. It sometimes has a dark stripe along the centre of its head. The fur is grey-brown, often grizzled with long dark hairs above, and grey-white below. The feet are slender and covered with white hairs. It has one pair of upper and lower gnawing incisor teeth and 3 pairs of molars on each side of the jaw. **Behaviour** Active mainly at night, New Holland mice live in family groups, sharing a permanent nest in a deep burrow up to 5 m long, with a nest chamber and a vertical entrance shaft concealed by shrubs or dead leaves. They also build short, shallow burrows, and use them as refuges to escape predators. They maintain separate home ranges, although those of adult females may overlap. Up to 17 animals per hectare may be found in sandy heaths with shrubs 2–8 years old, actively regenerating after fire. **Development** Females are sexually mature at 13 weeks, males at 20 weeks, and females can produce their first litter in the year of birth. The breeding season extends from August to January, and females may raise up to 4 litters of 2–6 (usually 4–5) young per year. They are born 32–39 days after mating, open their eyes at 15 days, are weaned at 3–4 weeks and have a lifespan of 18–24 months. **Diet** Seeds, flowers, fungi, moss, roots, leaves and insects. **Habitat** Dry coastal heaths, scrubs, woodlands and open forests with a dense understorey, particularly in areas regenerating after fire. **Traces** Scats are deposited at burrow entrances. They are brown to dark brown, 1–2 mm across and 2–5 mm long. **Status** Vulnerable nationally, but critically endangered in Vic. and endangered in Tasmania. Threatened by habitat destruction and predation by introduced feral animals.

WEIGHT **10–24 g** HEAD–BODY **65–95 mm** TAIL **79–110 mm**

FAMILY **MURIDAE** SPECIES *Zyzomys argurus*

COMMON ROCK-RAT

A compact rodent the size of a large mouse, the common rock-rat is found in rocky terrain with suitable nesting sites. It has coarse fur, golden-brown above and white below. Its sparsely-haired tail has overlapping scales and is swollen at the base with fat deposits. The skin of the tail is easily pulled off, and the tail withers away to a stump, hence many individuals have shortened or missing tails. The soles of the feet are smooth, the eyes are large and bulging, the ears rounded, and the muzzle has long whiskers. It has one pair of upper and lower gnawing incisor teeth and 3 pairs of molars on each side of the jaw. **Behaviour** Active at night, common rock-rats are very susceptible to heat stress and quickly die if exposed to the hot sun. They sleep in nests in cool refuges in rock crevices. Little is known about their behaviour. They fight ferociously when first caged together and may establish dominance hierarchies in the wild. Few are seen towards the end of the wet season when they probably form breeding pairs. **Development** Sexually mature at 5–6 months, they may live to 4 years in captivity, and breed year round with a slight lull in the wet season and a peak in April. Females have 4 teats and rear 1–4 young, born about 35 days after mating. The young are left in the nest while the mother forages. They are fully-furred at 10 days, have their eyes open at 12 days and are independent at about 4 weeks. **Diet** Plant stems, leaves, seeds, grasses, fungi and insects. **Habitat** Woodlands, low open forests and grasslands, always in rocky areas, particularly sandstone country. **Traces** Scats are yellow to brown pellets, 2–3 mm across and up to about 8 mm long, containing fine particles of plant matter. **Status** Low risk.

WEIGHT **26–65 g** HEAD–BODY **85–140 mm** TAIL **90–125 mm**

FAMILY **MURIDAE** SPECIES *Hydromys chrysogaster*

WATER-RAT

A rabbit-sized placental mammal, the water-rat is one of the largest Australian rodents. It has soft, dense, water-repellent fur varying from dark grey to black or brown above, and white, cream or golden-yellow below, with a white-tipped tail. The body is streamlined for swimming with a flattened head, long blunt muzzle, small eyes and ears and a thick tail used as a rudder. The hind feet are large and broad with webbing between the toes. They have one pair of upper and lower gnawing incisor teeth, but unlike most other rodents have only 2 molars on each side of the lower jaw. **Behaviour** Active mainly at night, water-rats also forage in the early morning and evening. They sleep in a grass-lined nest in a burrow dug into the bank of a waterway, or sometimes in a log. Burrows have a round entrance about 150 mm across, hidden under a log or in vegetation. They run parallel to the bank, and have one or more inner nest chamber some 200 mm high. Water-rats are solitary except when mating or with young, and fight to establish dominance hierarchies and defend territories of 2–10 ha or larger. Sluggish on land, they follow tracks along the water's edge to feeding sites, stalk through vegetation for prey and dive into the water around likely food sources, often emerging some distance away, carrying their catch to a favourite feeding platform. **Development** Sexually mature at 4–8 months, they continue to grow throughout their life of 3–4 years. Breeding occurs year-round with peaks in late winter and spring, and most young are born from September to February. They rear 1–5 litters annually of usually 3–4 young, born 34 days after mating. Females have 4 teats and suckle their young in the nest for about 4 weeks. They are independent at about 8 weeks and moult twice a year. **Diet** Fish, crustaceans, molluscs, frogs, water birds, eggs, bats and aquatic insects. **Habitat** Close to permanent fresh, brackish or saltwater wetlands, creeks, rivers and estuaries, with good vegetative cover; to about 1500 m. **Traces** Scats, shells and other remains near feeding platforms. Scats are up to 10 mm across and 20–25 mm long, containing coarse fragments of shells, scales and bones. **Status** Low risk.

WEIGHT **340–1280 g** HEAD–BODY **230–370 mm** TAIL **225–345 mm**

FAMILY **MURIDAE** SPECIES *Melomys burtoni*

GRASSLAND MELOMYS

A rat-sized placental mammal, the grassland melomys is considered a pest by sugar cane farmers who have created new habitats for it by clearing forests. The fur is long and soft; grey-brown to reddish-brown or khaki above; sometimes pale orange on the flanks; and white, grey or cream below. It has a thick-necked appearance with short rounded ears, a broad head, bulging eyes, long whiskers and broad hind feet. Its slender tail is partially prehensile, dark brown and naked with small non-overlapping scales forming a mosaic tile pattern rather than rings. **Behaviour** Active at night, grassland melomys are agile climbers, foraging in tall reeds, sedges and sugar cane, using their partly prehensile tail for balance and grip. They can also cross waterways by swimming or rafting. They sleep by day in spherical nests of shredded leaves, bark and grass, 200–300 mm diameter, often with 2 entrances and usually about a metre above ground. Nests are woven around stout grass stems or sugar cane leaves, built in shrubs, pandanus trees, hollow logs, under sheets of bark, and sometimes in short burrows. Males occupy home ranges of about 0.4 ha, females about 0.2 ha, and they become very aggressive if confined. **Development** Grassland melomys breed throughout the year with peaks in the wet season in the Kimberley, in autumn and winter in Qld, and in spring and summer in the south (depending on the rainfall). Females have 4 teats and usually rear 2–3 young. Newborn develop rapidly and cling to the mother's teats until fully furred with their eyes open at 7–10 days old. They are carried by the mother clinging to her teats if she changes nests and are weaned at about 20 days. They have a lifespan of 1–2 years. **Diet** Grass stems, seeds, berries, sugar cane and insects. **Habitat** Grasslands, sedgelands, heathlands, open forests, woodlands, vine thickets, monsoon forests, mangroves, grassy clearings in rainforests, and canefields. **Traces** Scats are pale to dark brown cylindrical pellets about 2 mm across and up to 10 mm long, containing powdery plant material. **Status** Low risk.

WEIGHT **25–124 g** HEAD–BODY **125–160 mm** TAIL **90–175 mm**

FAMILY **MURIDAE** SPECIES *Melomys cervinipes*

FAWN-FOOTED MELOMYS

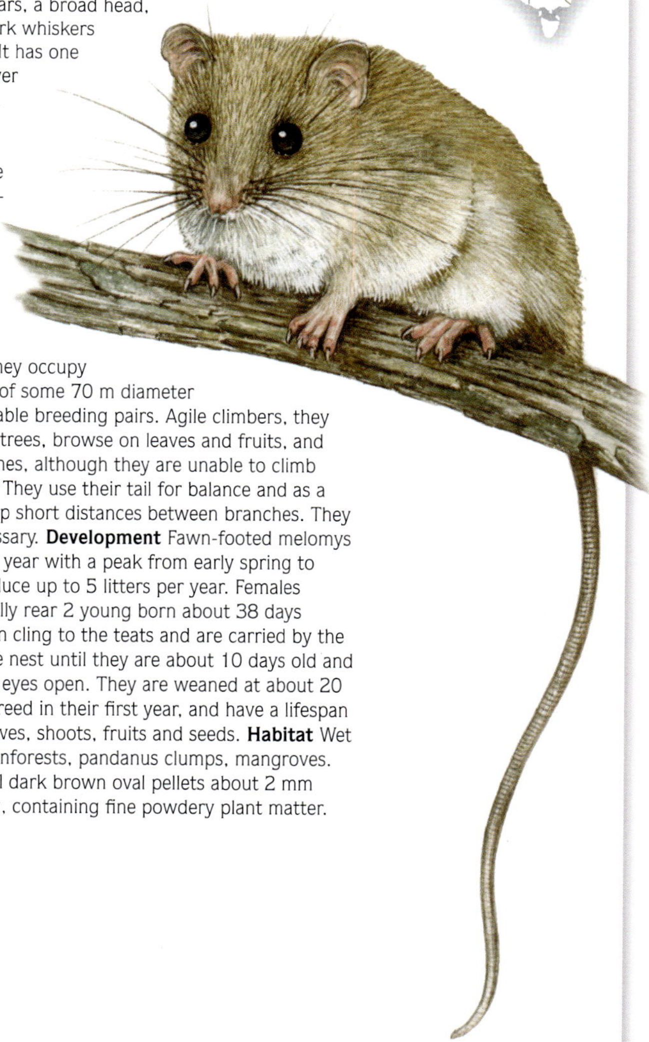

A rat-size placental mammal of the closed forests of eastern Australia, the fawn-footed melomys has long, soft, fine fur; sandy-brown to dark grey-brown or occasionally bright russet above; and white, cream or grey below. The naked tail is brown to black, partially prehensile with non-overlapping scales. It has a thick-necked appearance with short rounded ears, a broad head, bulging eyes, long dark whiskers and broad hind feet. It has one pair of upper and lower gnawing incisor teeth and 3 pairs of molars on each side of the jaw. **Behaviour** Active mainly at night, fawn-footed melomys sleep by day in grass-lined nests in trees, shrubs and pandanus fronds, and sometimes shelter in burrows. They occupy distinct home ranges of some 70 m diameter and probably form stable breeding pairs. Agile climbers, they forage in and around trees, browse on leaves and fruits, and climb creepers and vines, although they are unable to climb smooth-barked trees. They use their tail for balance and as a fifth limb, and can leap short distances between branches. They can also swim if necessary. **Development** Fawn-footed melomys breed throughout the year with a peak from early spring to early winter, and produce up to 5 litters per year. Females have 4 teats and usually rear 2 young born about 38 days after mating. Newborn cling to the teats and are carried by the mother if she flees the nest until they are about 10 days old and fully furred with their eyes open. They are weaned at about 20 days. Some females breed in their first year, and have a lifespan of 2.5 years. **Diet** Leaves, shoots, fruits and seeds. **Habitat** Wet sclerophyll forests, rainforests, pandanus clumps, mangroves. **Traces** Scats are small dark brown oval pellets about 2 mm across and 5 mm long, containing fine powdery plant matter. **Status** Low risk.

WEIGHT **45–115 g** HEAD–BODY **95–200 mm** TAIL **115–210 mm**

FAMILY **MURIDAE** SPECIES *Uromys caudimaculatus*

GIANT WHITE-TAILED RAT

A rabbit-sized placental mammal, the giant white-tailed rat is one of Australia's largest rodents. It has coarse fur with long spiny guard hairs, grey-brown above and creamy-white below, with pale paws. The tail is almost naked with non-overlapping scales; the terminal third is white, the remainder is dark grey mottled with white patches. The muzzle is long with many black whiskers. The eyes and ears are small. The limbs are short with large, strongly-clawed hind feet. It has one pair of upper and lower gnawing incisor teeth and 3 pairs of molars on each side of the jaw. **Behaviour** Active mainly at night, giant white-tailed rats sleep by day in tree hollows or burrows dug beneath logs or in stream banks, lined with vegetation. They are found alone or in pairs, and have overlapping home ranges of 4 ha or more. Males are very aggressive to other males during the breeding season and emit harsh threatening growls. Fast and agile climbers, they forage on the ground and in trees, using their semi-prehensile tail to grip branches, bounding up trees, gripping with their forefeet and pushing with their strong hind legs. Their jaws are able to break hard-shelled nuts, which are often taken to a favourite feeding spot on a log or tree buttress. Some seeds are buried near the tree, marked with a leaf and eaten the following night. Beetles are dug from rotting logs and bark is also eaten. **Development** Sexually mature at about 10 months, they breed from September to January. Females have 4 teats and rear 2–4 young born 36 days after mating. Newborn cling to the teats and may be carried by the mother if she leaves the nest. They suckle for about 5 weeks, are independent at about 2–3 months and may live to 4 years or more. **Diet** Nuts (including coconuts), fruits, insects, fungi, bark, small reptiles, amphibians, crabs and eggs. **Habitat** Wet sclerophyll forests, rainforests, woodlands, mangroves and melaleuca forests. **Traces** Chewed seed cases at feeding sites; bark torn from buttressed trees. Scats are dark brown or black, 3–4 mm across and 10–15 mm long, containing fine powdery particles and fragments of insects and scales. **Status** Low risk.

WEIGHT **500–890 g** HEAD–BODY **275–385 mm** TAIL **320–362 mm**

FAMILY **MURIDAE** SPECIES *Mus musculus*

HOUSE MOUSE

One of the most adaptable placental mammals in the world, the house mouse probably originated in central Asia and accompanied humans to Australia. The fur is soft and dense, to 7 mm long, brownish-grey above and white to grey or pale yellow below. The ears are large and rounded, the eyes bulging, and the tail sparsely-haired with overlapping scales. It can be distinguished from other small rodents by its distinctive musty smell and a notch on the inner surface of the upper incisor teeth. **Behaviour** Active mainly at night, house mice sleep in spherical nests of shredded material in secluded sites in buildings, reed beds, cracks in the ground, or at the centre of complex shallow burrow systems with several narrow tunnels and entrances. In buildings they live in small family groups and defend a fixed territory about 6 m diameter, although juveniles are forced to disperse. Less defined territorial behaviour is exhibited in the wild. Movements are unhurried, although they can move quickly and climb easily using their sharp claws. They are poor swimmers and become torpid in cold conditions. **Development** Sexually mature at 6–8 weeks, they usually breed from October to April, but in good years they can breed year-round. Females have 10 teats and may rear up to 11 litters of 4–10 young per year. Females mate soon after giving birth, the embryo remaining dormant for up to 18 days if she is lactating. Pregnancy lasts 19–21 days, infants open their eyes at 12 days and are weaned at 18 days. **Diet** Seeds, fruits, fungi, insects and household scraps. **Habitat** Most habitats, from deserts to wet coastal areas, but is displaced by small native mammals. **Traces** Distinctive musty odour. Scats are 1–2 mm across and 3–5 mm long, musty smelling, and often deposited in dark corners. **Status** Low risk.

WEIGHT **8–25 g** HEAD–BODY **60–100 mm** TAIL **75–100 mm**

FAMILY **MURIDAE** SPECIES *Rattus colletti*

DUSKY RAT

The dusky rat lives on the flat, treeless alluvial floodplains of tropical northern Australia and undergoes dramatic population fluctuations. It has long, coarse, spiny fur, grizzled dark brown to black above, yellowish on the sides and throat, and greyish-buff below. The small ears and long, stout, hairy tail are dark brown. Like other rodents it has one pair of upper and lower gnawing incisor teeth and 3 pairs of molars on each side of the jaw. **Behaviour** Dusky rats modify their lifestyle according to the dramatic seasonal variations of the wet tropics. In the wet season they retreat from rising floodwaters to higher ground and forage at night in shallow water or in the woodlands, sheltering by day beneath tree roots or logs, or in shallow burrows dug into soft soil. As the floodwaters recede they disperse over the drying plains and spend most of their time in crevices in the clay pans foraging for underground sedge corms. Breeding begins at the end of the wet season and the young are born in nests constructed by the mother in humid crevices. As the dry season progresses they take refuge from the heat in cracks and crevices, and above-ground nocturnal activity stops completely. At this time of year the older animals die. When the rains begin and the cracks in the soil vanish they move out once more to the edge of the plains. **Development** Females are able to breed at about 5 weeks of age. They have 12 teats and give birth to usually 7–9 young, 21–23 days after mating, and wean them at about 20 days. If there is only mild flooding in the wet season, and rain extends into the dry season, they will breed throughout the year, although little reproduction takes place in years with extreme weather conditions. **Diet** Grass roots and sedge corms. **Habitat** Coastal floodplains, swamps and tidal rivers fringed with mangroves. **Traces** Scats are pale brown, 2–4 mm wide and about 10 mm long, often curved, containing fine powdery particles.
Status Low risk.

WEIGHT **85–215 g** HEAD–BODY **145–210 mm** TAIL **105–150 mm**

FAMILY **MURIDAE** SPECIES *Rattus fuscipes*

BUSH RAT

Widespread and abundant in the forest understorey of the coast and adjacent ranges, the bush rat has dense soft fur, grey to grey-brown or slightly reddish above, and grey or cream below. The tail is grey to brown, sparsely-haired with overlapping scales. The soles of the hind feet are pale, distinguishing it from the swamp rat which has black soles. It has one pair of upper and lower gnawing incisor teeth and 3 pairs of molars on each side of the jaw. Males are much larger than females. **Behaviour** Active at night, bush rats shelter by day in rock crevices, beneath fallen timber or in burrows 2–3 m long, often built under logs or stones. Burrows have a sloping, twisting tunnel leading to a vegetation-lined nest chamber some 150 mm diameter and 350–450 mm below ground. Adults live alone or with their young, and defend an exclusive home range of around 0.1–0.4 ha, although males roam over large areas in spring looking for females. Juveniles disperse and must establish a small home range with access to sufficient resources to survive the winter. **Development** Bush rats are sexually mature at 3–4 months. They can breed throughout the year, with birth peaks in late spring and summer, and may produce up to 4 litters in a good year. Females have 10 teats (sometimes 8 in the north) and give birth 20–21 days after mating. Litters usually comprise 4–5 young weighing about 5 g each. Their eyes are open by 3 weeks and they are independent at 4–5 weeks. They have a lifespan of up to 3 years, although in cool climates few survive their first winter. **Diet** Insects, supplemented by fungi, leaves, fruit and seeds. **Habitat** Rainforests, heaths, sedgelands and wet sclerophyll forests with good ground cover of shrubs and ferns; to more than 1800 m. **Traces** Seed husks cracked lengthways. Scats are dark brown, about 2–3 mm across and 5–15 mm long, with a pungent odour. **Status** Low risk.

WEIGHT **40–225 g** HEAD–BODY **110–215 mm** TAIL **105–195 mm**

FAMILY **MURIDAE** SPECIES *Rattus leucopus*

CAPE YORK RAT

A rainforest rodent found in Cape York and New Guinea, the Cape York rat is divided into two distinct Australian populations separated by a wide dry corridor around Laura. It has coarse spiny fur, blackish-brown to golden-brown or reddish above and white or grey below. The head is pointed with large ears, sparsely covered with buff to reddish-brown hairs. The tail is slender and tapering, almost hairless, often mottled brown or grey, and has off-white patches in the most northerly Australian population. The pale hind feet are stout and long. It has one pair of upper and lower gnawing incisor teeth and 3 pairs of molars on each side of the jaw. **Behaviour** Active mainly at night, Cape York rats are ground-dwellers, foraging on the forest floor among the leaf litter and debris for insects and other foods. They shelter by day in burrows usually with several entrances hidden beneath the leaf litter, and a number of interconnecting chambers lined with dry leaves. Holes in tree bases and rotting logs are also used for shelter. Individuals may use more than one burrow and some burrows appear to be communal. **Development** Females are sexually mature at about 3 months. Cape York rats breed throughout the year with a lull in the dry season from June to early September, and can produce up to 3 litters per year. Females have 6 (rarely 8) teats and usually give birth to 2–5 young, 21–24 days after mating. Their eyes are open by 22 days and they are weaned at about 25 days. **Diet** Insects, supplemented by fungi, herbs, seeds and fruit. **Habitat** Tropical rainforests. **Status** Low risk.

WEIGHT **70–215 g** HEAD–BODY **130–215 mm** TAIL **135–210 mm**

FAMILY **MURIDAE** SPECIES *Rattus lutreolus*

SWAMP RAT

A robust, cryptic rodent, the swamp rat has long soft fur, dark grey to grey-brown above and lighter-grey or buff below. The tail is short, sparsely-haired with overlapping scales. The ears are short and largely concealed by the fur. The feet are dark grey-brown, and the hind feet are relatively short. It has one pair of upper and lower gnawing incisor teeth and 3 pairs of molars on each side of the jaw. **Behaviour** Although mainly nocturnal, swamp rats are frequently active during the day and sleep in spherical nests of shredded vegetation about 150 mm across, built at the end of a burrow up to 1 m deep, or above ground in tussock grass, rotting logs or hollows in tree bases. Solitary, they are strongly territorial and have small home ranges of about 0.2–0.5 ha. Males increase their home ranges to about 4 ha in the breeding season and roam widely through adjacent female territories searching for mates. Young adults must disperse more than 2 km to establish a new territory. To avoid predators they construct extensive tunnels and runway systems through dense vegetation from their nesting sites to food sources. They can climb and swim fast in emergencies. **Development** Females are sexually mature at about 12 weeks and may breed in their first year. Breeding occurs mostly in early spring and summer, and several litters of 3–5 young are raised in a good season. Mainland females have 10 teats, while those in Tasmania have only 8 teats. Young are born 21–23 days after mating, are weaned at 3–4 weeks and live to about 18 months. **Diet** Rhizomes, stems and leaves of sedges and grasses; supplemented by insects, mosses, seeds and fungi. **Habitat** Coastal swamps, river flats, wet heaths, button grass moorlands, sedgelands, grasslands, woodlands and sclerophyll forests with dense, low vegetation; to more than 1000 m.
Traces Scats are found along runways and nesting sites. They are brown, about 2–5 mm across and 10–15 mm long.
Status Low risk.

WEIGHT **55–170 g** HEAD–BODY **120–200 mm** TAIL **55–150 mm**

FAMILY **MURIDAE** SPECIES *Rattus norvegicus*

BROWN RAT

Introduced from Europe, the brown rat is a thickset placental mammal with shaggy, bristly fur, grey-brown above and white to grey below. The ears are small and rounded, close set and up to 20 mm long. The muzzle is blunt with long whiskers. The tail is stout, paler below, short and naked with overlapping scales. The pink hind feet have long toes and pale soles. Like other rodents they have one pair of upper and lower gnawing incisor teeth and 3 pairs of molars on each side of the jaw. **Behaviour** Active mainly at night, brown rats can climb and swim well, but prefer to live at ground level, following odour trails along runways to food sources and nest sites. They shelter under bushes, rocks, in sewers and other urban sites, or construct extensive burrow systems, huddling together in nests of shredded material carried in the mouth. They hoard food in the nest, carrying it in their cheeks and manipulating it with their forefeet. Brown rats often live in colonies of several hundred, recognising each other by smell and maintaining individual home ranges about 30 m across. Dominant males have exclusive territories around burrows containing several breeding females, and juveniles are forced to leave the colony. **Development** Sexually mature at 6–8 weeks, they are able to breed throughout the year, producing up to 5 litters per year. Females have 12 teats and give birth to 6–11 young, 21–23 days after mating. The young are well-furred at 10 days, open their eyes at about 15 days and can be weaned at 20 days. They seldom live more than 12 months. **Diet** Most plant and animal material. They kill and scavenge small mammals and nestlings, and eat birds eggs. **Habitat** Degraded stream banks, rocky beaches, sewers, farm buildings and warehouses. **Traces** Spoil heaps at burrow entrances, nests under floors, gnawed seed fragments and greasy rub-marks beside runways. Scats are brown, about 3–5 mm across and 10–20 mm long. **Status** Low risk.

WEIGHT **200–480 g** HEAD-BODY **180–260 mm** TAIL **150–215 mm**

FAMILY **MURIDAE** SPECIES *Rattus rattus*

BLACK RAT

Introduced from Europe, the black rat has become well established in coastal Australia. It is a slender gentle rodent with a sleek smooth coat, charcoal-grey to black or light brown above, and cream or white below. The head is long and pointed with large thin ears more than 20 mm long, protruding eyes and long black whiskers. The tail is long and naked with overlapping scales. Like other rodents they have one pair of upper and lower gnawing incisor teeth and 3 pairs of molars on each side of the jaw. They may carry the plague bacillus and transmit other diseases to humans through their excreta. **Behaviour** Active mainly at night, black rats are agile climbers and good swimmers. They shelter in nests of shredded material in roofs, cavity walls, trees, shallow scrapes, or extensive shallow burrow systems, often around farm buildings or river banks. They mark their territory with scent by rubbing with the cheek or belly. Although timid when cornered, they will defend their territory against intruders. Territorial groups establish a hierarchy with a dominant resident male and subordinates. Mated pairs share a nest until shortly before the female gives birth, when she excludes him. If disturbed the female moves her litter to a new nest. **Development** Sexually mature at 3–4 months, they live to about 12 months in the wild and breed throughout the year, producing up to 6 litters. Females have 10–12 teats and usually give birth to 4–10 young, 21–22 days after mating. Young are born blind and naked, and are weaned at about 28 days. **Diet** Grass stems and roots, seeds, insects, underground fungi, small mammals, birds and eggs. **Habitat** Disturbed areas, watercourses, urban areas and in buildings. **Traces** Fine, gnawed seed fragments and greasy smears where they brush past walls. Scats are usually dark brown, 2–3 mm across and 5–15 mm long. **Status** Low risk.

WEIGHT **95–340 g** HEAD–BODY **160–210 mm** TAIL **180–255 mm**

FAMILY **MURIDAE** SPECIES *Rattus sordidus*

CANEFIELD RAT

The canefield rat is a rodent of the tropical grasslands and open forests, and is regarded as a serious pest by sugar cane growers who have created an ideal habitat for it. A short-limbed placental mammal, the canefield rat has coarse spiny fur grizzled dark brown to black above with prominent guard hairs, and light grey or buff belly fur. The tail is dark brown to black with prominent overlapping scales. The soles of the hind feet are pale, distinguishing it from the swamp rat, which has dark soles. It has one pair of upper and lower gnawing incisor teeth and 3 pairs of molars on each side of the jaw. **Behaviour** Canefield rats are usually nocturnal, and forage in dense grass or sugar cane plantations with a complete canopy of leaves. When population densities are high they are active during daylight hours. They rest in extensive burrow systems about 400 mm deep, with tunnels 50–100 mm diameter, sloping down to a nest chamber about 150 mm diameter lined with dry grass. Burrow entrances are near grass clumps, stones and stumps, and burrows are often concentrated in dense colonies with networks of runways between them. As many as 23 non-breeding adults have been found in one nest chamber, although females with young force other adults out. **Development** Females are sexually mature at 9–10 weeks, males at about 5 months, with a lifespan of 2–3 years. Breeding occurs at any time of year, with the majority of births from March to May. Females have 12 teats and usually give birth to 6 young, 20–27 days after mating. Young are fully furred at 7 days, have their eyes open at 18 days, and are weaned at 3 weeks. Canefield rats have the highest reproductive potential of all the native rats, and may reach plague proportions in canefields. **Diet** Grass, seeds, leaves, sugar cane stems and insects. **Habitat** Tropical grasslands with dense ground cover, open forests, grassy clearings in rainforests, and canefields. **Traces** Scats are pale to dark brown, up to 5 mm across and 20 mm long, comprising powdery plant fragments. **Status** Low risk nationally, but vulnerable in the NT (South-West Island). Threatened by cat predation.

WEIGHT **50–260 g** HEAD-BODY **123–210 mm** TAIL **100–160 mm**

FAMILY **MURIDAE** SPECIES *Rattus tunneyi*

PALE FIELD-RAT

Usually found in tall grasslands, the pale field-rat is an attractive rodent with light shiny yellow-brown fur above and grey or cream fur below. The head is broad and rounded with protruding eyes and short rounded ears. The tail is relatively short and sparsely-haired with overlapping scales. It has one pair of upper and lower gnawing incisor teeth and 3 pairs of molars on each side of the jaw. **Behaviour** Pale field-rats are strictly nocturnal and spend the day alone or with their offspring in shallow burrow systems dug into loose sandy soil, or in termite mounds. Burrows have vertical shafts and horizontal tunnels, and often have several entrances marked by large spoil heaps. They appear to live in patchy colonies, and travel between burrows along well-marked runways through the thick grass. Pale field-rats are docile, curious and easily handled, and will enter houses searching for food. **Development** Females are sexually mature at about 5 weeks. Breeding takes place in spring in NSW and between January and August in the tropics. Females have 10 teats and may rear several litters of 2–10 (usually 4–5) young in quick succession. Pregnancy lasts 21–22 days. Young are furred at 7 days, have their eyes open at 18 days and are weaned at 21 days. **Diet** Grass roots, stems and seeds. **Habitat** Tall grasslands, mangroves, sand dunes, open forests with a grassy understorey, canefields and hoop pine plantations. **Traces** Spoil heaps and scats around burrow entrances. Scats are dark to light brown, up to 5 mm across and 15 mm long, containing fine powdery particles. **Status** Low risk nationally, but endangered in SA. Threatened by habitat destruction and predation by introduced feral animals.

WEIGHT **40–206 g** HEAD–BODY **120–198 mm** TAIL **78–190 mm**

FAMILY **MURIDAE** SPECIES *Rattus villosissimus*

LONG-HAIRED RAT

This desert rodent is also known as the plague rat because its numbers reach plague proportions after several continuous wet years. The fur is long and grey above with conspicuous black guard hairs, and light grey below. The tail is grey to black with prominent dark hairs and overlapping scales. It has one pair of upper and lower gnawing incisor teeth and 3 pairs of molars on each side of the jaw. **Behaviour** Active mainly at night, long-haired rats shelter by day in burrows dug under bushes or rocks. Adult males are solitary and use several shallow, temporary burrows. Females live alone or with their young and construct complex breeding burrows to about 1 m deep and 3 m long. These burrows have several side passages and a central nesting chamber 100–150 mm diameter, containing a spherical nest of shredded vegetation. They are grouped in grassy depressions with several well-worn runways between them. In plague years these rats wander in vast numbers over the countryside, invading camps and homesteads in their thousands before suddenly dying out after a few weeks. **Development** Females are sexually mature at about 10 weeks and have a very high reproductive potential. They breed throughout the year with lulls from mid-winter to early spring. Females have 12 teats and give birth to usually 6–7 blind young, born 21–23 days after mating. The young open their eyes at 17 days, are weaned at about 3 weeks and have a lifespan of about 18 months. **Diet** Grasses, herbs, seeds, flowers and insects. In plagues they become cannibalistic and will attempt to eat any organic material. They need water and retreat to moist areas during droughts. **Habitat** Most habitats in arid to semi-arid areas with access to drinking water.
Traces Spoil heaps at burrow entrances.
Status Low risk nationally, but vulnerable in NSW. Threatened by disturbance of burrows by stock.

WEIGHT **105–290 g** HEAD-BODY **120–225 mm** TAIL **100–190 mm**

FAMILY **CANIDAE** SPECIES *Canis lupus dingo*

DINGO

Australia's largest native carnivore, the dingo was introduced 3500–4000 years ago. A form of the domestic dog, it is related to Asian and Middle Eastern wolves. It is typically ginger to sandy-brown with white points, although some are black with sandy-brown markings. The ears are held erect, the snout is narrower than that of the domestic dog, while the canine teeth are larger and more slender. They have a bushy tail with a scent gland at the base. The legs are slender with 5 toes on the forefeet and 4 on the hind feet. The claws are straight and non-retractable. They interbreed with domestic dogs, and wild populations often include hybrids. **Behaviour** Dingos are active mainly around dawn and dusk, with bursts of activity throughout the day and night. They shelter in caves or in the shade, and females rear their young in a den in a cave, hollow log, enlarged wombat burrow or in thick vegetation. Although they act independently, dingos belong to loose associations, distinguished by scent-marking. They hunt alone for small game and cooperate to hunt larger game such as kangaroos, forming packs with a dominant male and female and 4–5 subordinates (usually offspring of the dominant pair). They rarely fight, maintaining hierarchies by ritualised postures, and usually remain in a home range up to 80 sq km. Short term associations are formed during the breeding season. Non-breeding yearlings often help to rear the young, standing guard and coaching them in hunting techniques. Dingos howl to keep contact when hunting and to attract mates. A howl-bark is used as an alarm signal. **Development** Sexually mature at 12 months, dingos breed once a year, giving birth to 2–9 pups from March to September, 63 days after mating. Newborn are blind and well-furred, and are mobile by about 4 weeks. They suckle for 3–4 months and often follow their parents until 12 months old. They have a lifespan of up to 14 years. **Diet** Mammals, reptiles, birds and insects. They can survive for long periods without drinking. **Habitat** Most habitats from alpine forests to tropical rainforests and arid areas with access to drinking water. **Traces** Bones and feathers near den entrance. Scats are often deposited on rocks and grass tussocks. They are up to 40 mm across and 100 mm long, often twisted with tufts of hair at one end, and contain bone fragments. **Status** Low risk.

WEIGHT **9.5–24 kg** HEAD–BODY **860–1220 mm** TAIL **260–380 mm**

FAMILY **OTARIIDAE** SPECIES *Neophoca cinerea*

AUSTRALIAN SEA LION

The only sea lion endemic to Australia, this marine mammal has a bulky, streamlined body. It has a blunt snout and a dog-like head, with long whiskers and very small, rolled, external ears. Adult males are rich chocolate-brown to blackish with a long mane of white hair over the neck. Adult females and young are silver-grey to fawn above, and yellow to cream below. They have flippers instead of forelimbs. The hind legs are webbed and face the rear. Males are much larger than females. **Behaviour** Active by day, Australian sea lions forage along the edge of the continental shelf, generally staying within 300 km of their birthplace. They are powerful swimmers, diving, leaping out of the water and surfing onto beaches, obtaining most thrust from their foreflippers. They can also gallop and climb cliffs, and males have been found 10 km inland. Sea journeys tend to be short and throughout the year they congregate at traditional breeding grounds. Some colonies comprise more than 500 animals, but the majority are much smaller. Numbers peak during and after the pupping season, when males gather females into a territory, fighting and threatening other males. The strongest males maintain the largest harems. **Development** Females are sexually mature at about 3 years, males at 6 years, with a lifespan of 12 years or more. Females breed every 17–18 months, producing a single pup from October to January. They mate 4–9 days after giving birth, although the embryo remains dormant for about 4 months and takes another 14–15 months to develop. Newborn are 600–750 mm long and weigh 6–8 kg. They have thick, soft fur, moult at 2 months, swim at 3 months and suckle for 12–18 months. The mother stays with the pup for about 2 weeks then leaves it in a sheltered place while she feeds at sea, returning every 2 days or so. **Diet** Squid, rock lobsters, fish and sea birds. **Habitat** Cool temperate coastal waters, isolated sandy beaches and rocky coastlines. **Status** Vulnerable in WA and rare in SA.

WEIGHT **60–300 kg** LENGTH **1.3–2.5 m**

FAMILY **OTARIIDAE** SPECIES *Arctocephalus forsteri*

NEW ZEALAND FUR SEAL

The New Zealand fur seal is a streamlined marine mammal with a dog-like head, pointed snout with long whiskers and small, rolled, external ears. It is dark grey-brown above and lighter below. Bulls have a massive neck and a thick mane of coarse hair. Newborn are black and moult at 2 months to become silver-grey. They have strong flippers and webbed hind legs facing the rear. Males are much larger than females. **Behaviour** New Zealand fur seals rest, moult and breed on land, where they move with a slow shuffling walk. They are powerful swimmers, propelling themselves with vertical strokes of their webbed hind limbs. They feed mostly at night, diving 12–18 m (up to 163 m) and staying underwater for up to 6 minutes. Breeding colonies of up to 1300 begin to form in October when bulls join cows and juveniles on exposed shores of offshore islands around rock slopes, ledges, boulder-strewn beaches and caves. Dominant bulls establish a territory among the females by posturing, guttural barking or savage fighting, and defend a harem of about 6 cows. Unsuccessful bulls and young males congregate on the edge of the breeding area. Bulls leave in late January and swim off to hauling grounds on rocky beaches. **Development** Sexually mature at 4–6 years, males only breed when strong enough to defend a territory at 8–10 years of age. Births occur from November to January and the cows mate 8 days after giving birth to a single pup. The embryo remains dormant for 3–4 months and takes another 8–9 months to develop. The newborn are about 450 mm long and weigh some 4 kg. The mother goes to sea to feed 10 days after giving birth, returning 3–5 days later to suckle for 2–4 days before leaving again. Pups congregate in pods and are suckled for 9–10 months. They have a lifespan of up to 17 years.

Diet Squid, fish, octopus, rock lobster, crab and some seabirds. **Habitat** Cool temperate coastal waters and offshore islands with rocky beaches. **Status** Low risk nationally, but rare in Tasmania.

WEIGHT **35–200 kg** LENGTH **1–2.5 m**

FAMILY **OTARIIDAE** SPECIES *Arctocephalus pusillus*

AUSTRALIAN FUR SEAL

The largest and most abundant seal in Australia, this streamlined marine mammal has a dog-like head with long whiskers and small, rolled external ears. Bulls are dark grey-brown with a mane of long coarse hair. Cows are silver-grey with a creamy-yellow throat and chest, and a chocolate-brown abdomen. Pups are dark brown above, yellowish below and moult at 3 months. They have flippers for forelimbs and webbed hind legs facing the rear. Males are much larger than females. **Behaviour** Australian fur seals spend most of the time at sea, returning to rest, moult and breed on land, where they move with a slow shuffling walk or a fast gallop. They swim with vertical strokes of their webbed hind limbs and can dive to at least 200 m. They have good underwater vision but poor vision on land, and may use echo-location underwater. Breeding grounds are occupied year-round by females and juveniles. Bulls arrive to breed in late October and the most powerful reoccupy territories by roaring, barking and fighting off intruding bulls. Females come ashore slightly later and fight each other for smaller territories within a male's territory, becoming part of his harem of up to 50 cows. Unsuccessful bulls and bachelor males congregate on the periphery. Males disperse to feed at sea in January. **Development** Sexually mature at 4–5 years, males only breed when they can defend a territory at 8–13 years old (until defeated 3–6 years later). Females begin breeding at 3–6 years, give birth from October to late December and mate 5–7 days later. The embryo remains dormant for 3–4 months and takes another 9 months to develop. Newborn are 600–700 mm long and weigh 4–13 kg. Pups congregate in pods and the cows go to sea after mating, returning weekly to suckle their pups for a further 8 months. The pups then accompany the cow at sea until weaned by 12 months. They have a lifespan of 19–21 years. **Diet** Squid, fish, octopus and rock lobsters. **Habitat** Cool temperate coastal waters, rocky coastlines. **Status** Low risk nationally, but rare in SA and vulnerable in NSW.

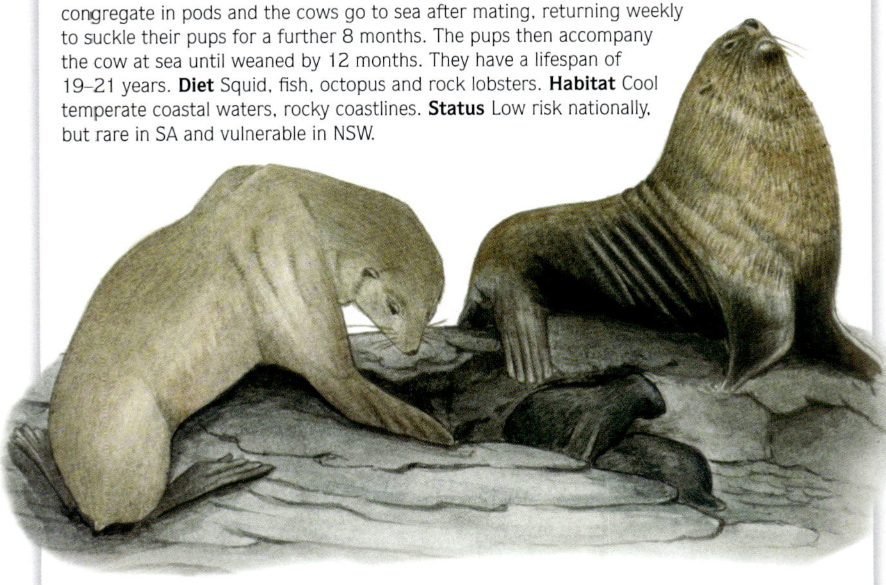

WEIGHT **36–360 kg** LENGTH **1.2–2.3 m**

FAMILY **PHOCIDAE** SPECIES *Hydrurga leptonyx*

LEOPARD SEAL

A summer inhabitant of the outer fringes of the Antarctic pack-ice, this streamlined marine mammal moves into Australian coastal waters in winter. It has a reptilian-like head with wide gaping jaws and a narrow neck. Its long, slim body is dark grey above and lighter below with grey or black spots on the sides and throat. Juveniles have a dark stripe on their back and dark spots below. It lacks an external ear and has distinctive 3–pronged cheek teeth. It has long broad flippers for forelimbs, and the webbed hind legs are turned backwards to act as tail flukes. **Behaviour** Leopard seals rest, moult and give birth out of the water, heaving their whole body forward in a caterpillar-like motion. They are fast and powerful swimmers, propelled by long, simultaneous strokes of their forelimbs as well as their hind limbs, and can dive for up to 15 minutes. They are generally solitary and form monogamous breeding pairs in summer. Young seals disperse north in winter, and some reach the coast of Australia between July and October, venturing as far north as Heron Island. They have a vocal repertoire of gurgles, grunts, chirps and whistles, and a throaty alarm call. **Development** Females are sexually mature at 3–7 years, males at 6–7 years. They mate in the water from November to February after weaning their previous young. The embryo remains dormant for some 3 months and takes another 7–8 months to develop. A single pup about 1.5 m long is born on the ice from October to January, and is suckled for about 4 weeks. They have a potential lifespan of about 26 years. **Diet** Krill, fish, cephalopods, seabirds and the young of other seals. They toss penguins around, biting off the flesh and leaving the skin. **Habitat** Pack ice, Antarctic coastline, subantarctic islands, cold to temperate coastal waters. **Status** Low risk nationally, but rare in SA.

WEIGHT **200–590 kg** LENGTH **2.4–3.8 m**

FAMILY **DELPHINIDAE** SPECIES *Delphinus delphis*

COMMON DOLPHIN

Commonly seen in Australian coastal waters, this torpedo-shaped marine mammal is dark grey to purple-black above, and cream to white below, with a distinctive hourglass pattern on its flanks. The tips of the snout, lower jaw and lips are purplish-black, and the eyes have black patches around them. It has a long slender beak, a slender, almost triangular dorsal fin, thin tapering flippers, and thin tail flukes with a slight notch in the edge. The jaws have 40–58 pairs of small teeth. **Behaviour** Sociable, friendly and playful, common dolphins move in groups of 5–50 and occasionally congregate in herds of 3000 or more, following the movements of schools of migrating fish. Small feeding groups scatter in the late afternoons to feed on food organisms rising to the surface before returning to the herd to rest or play. They can dive to 280 m, stay underwater for 8 minutes or more, and are often seen playing in the bow-waves of ships. They communicate over long distances using various pulsed sounds including squawks, squeaks, yaps and shrill whistles. In dark and murky waters they use echo-location to navigate and find prey. **Development** Females are sexually mature at about 8 years, males at about 9 years, and they have a lifespan of about 30 years. Common dolphins mate in spring and autumn at intervals of 1–3 years. A single calf, 700–900 cm long, is born 10–11 months later, and is fed on milk squirted by the mother into its mouth for about 18 months. **Diet** Fish and cephalopods, including herrings, anchovies, sardines and squid. **Habitat** Warm temperate, subtropical and tropical coastal waters, and deep offshore oceanic waters; generally between 10° and 28°C. **Status** Low risk.

WEIGHT **100–136 kg** LENGTH **1.7–2.4 m**

Family **Delphinidae** Species *Tursiops truncatus*

Bottlenose Dolphin

A coastal, torpedo-shaped marine mammal, the bottlenose dolphin is dark or light grey above, grading to light grey on the flanks, and white or pink below with a blue-grey band from the base of the beak to each eye. The head is robust with a relatively short beak with 18–26 pairs of teeth in each jaw. The lower jaw extends beyond the upper jaw and curves up slightly. The flippers are pointed, the dorsal fin is sickle-shaped, and the tail flukes are thin with rounded tips and a notched edge. Males are much larger than females. **Behaviour** Friendly and playful, bottlenose dolphins usually swim close to shore. They can remain submerged for several minutes, but usually rise to breathe every 15–20 seconds. They swim in pods of 5–20, forming part of a herd of several hundred, and communicate over long distances using pulsed sounds including squawks, squeaks, yaps and shrill whistles. Females whistle almost continually for several days after giving birth to give an acoustic imprint to their young. In dark and murky waters they use echo-location to navigate and detect prey, and if threatened will chase off and sometimes kill a shark. Herds have a home range of around 85 sq km, while individual home ranges vary from 15 to 40 sq km. Males fight over females in the breeding season and breeding females form relatively stable groups, caring for each other's infants while they feed. Subadult males associate in bachelor groups. **Development** Males are sexually mature at 10–14 years, females at 5–12 years, and they have a lifespan of 50 years or more. Females may produce 8 offspring during their lives, giving birth at 2–3 year intervals. Calves are born 12 months after mating, from February to May and September to November. Newborn are 900–1300 cm long, and feed on milk squirted from the mother's mammary glands for up to 6 years. **Diet** Fish, squid, eels and crustaceans. **Habitat** Warm temperate, subtropical and tropical coastal waters, bays and estuaries, occasionally ascending the tidal reaches of rivers. **Status** Low risk.

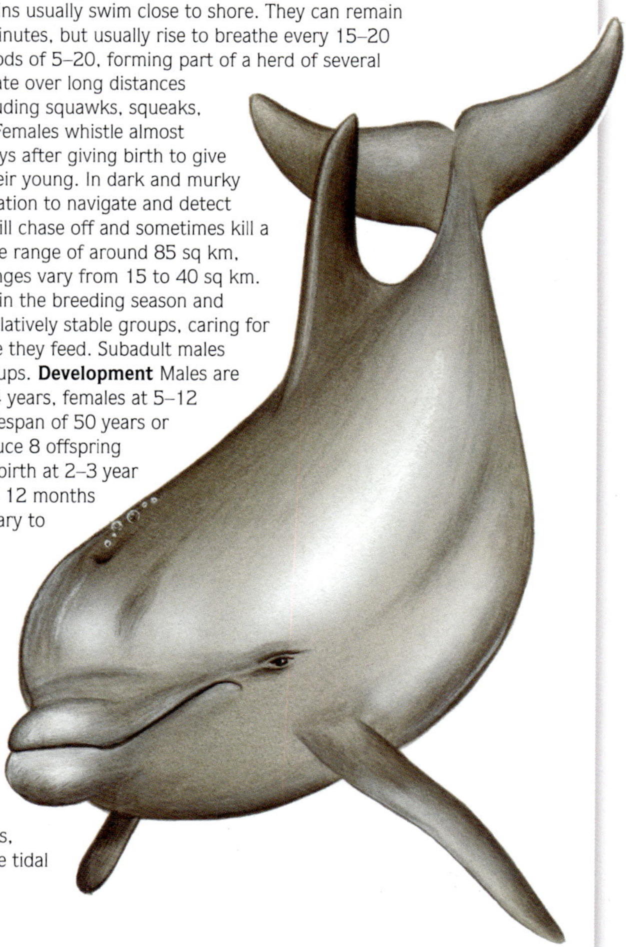

Weight **150–650 kg** Length **2.3–4 m**

FAMILY **DUGONGIDAE** SPECIES *Dugong dugon*

DUGONG

The dugong is the only herbivorous marine mammal. It has a bulky streamlined body, grey to bronze above and lighter below, with fine scattered hairs; older animals have large white patches. It has paddle-like forelimbs and a horizontal tail-fluke instead of hind limbs. It has a broad, flat head, small eyes, small ear-openings, a large mouth with fleshy lips, short bristles on the muzzle, and nostrils on the top of the head which close when diving. Males have a pair of protruding, tusk-like upper incisors. **Behaviour** Dugongs spend their lives in the water, feeding close to shore during daylight hours and travelling about 25 km daily. They dive repeatedly for food, staying underwater for up to 8.5 minutes (typically 1–3 minutes). They can swim at up to 22 kph with vertical strokes of their tail-flukes, using their flippers to change direction and brake. Active and alert, they live alone or in small family groups, sometimes congregating in herds of several hundred, and show no evidence of large-scale migration. Females give birth in shallow water to avoid sharks. They have good vision and hearing, and communicate with whistling and chirping sounds. **Development** Sexually mature at 9–15 years, they mate from May to November at 3–6 year intervals, giving birth to a single young 12–14 months later. Females have 2 teats in the armpits and males have internal testes. Calves weigh 20–35 kg, are 1–1.25 m long, and ride on the mother's back when not suckling, staying with her for about 2 years. Dugongs have a lifespan of up to 73 years. **Diet** Seagrasses supplemented with algae. **Habitat** Shallow tropical coastal waters and estuaries. **Status** Vulnerable nationally and endangered in NSW. Threatened by fishing activities, pollution and habitat destruction.

WEIGHT **250–900 kg** LENGTH **2.5–4 m**

Family **Balaenopteridae** Species *Megaptera novaeangliae*

Humpback Whale

A huge but friendly marine mammal, the humpback whale is grey-black above and black and white below, with a humpback and a broad, round, flattened head. Barnacle-encrusted knobs protrude from its jaw, and on each side of the roof of the mouth hang 270–400 fringed sheets of a horny material (baleen) used to sieve krill from the water. Its flippers are up to 16 m long and have knobs on their trailing edges; the dorsal fin varies from a small protuberance to a sickle-shaped fin. **Behaviour** Humpback whales can swim at 27 kph and dive 6–7 m deep for up to 30 minutes, surfacing for 3–6 minutes between dives, often leaping out of the water and rolling in mid-air. They use echo-location to navigate in murky waters, and migrate annually from their summer feeding grounds in the Southern Ocean to tropical waters to calve and mate, leaving in May and returning in November. One population migrates up Australia's west coast and another follows the east coast. Females with calves and immatures make the journey first, followed by adult males, then non-pregnant cows, with pregnant cows arriving last. Males advertise their arrival in the breeding area by singing a song of cries, yups and chirps, lasting 6–35 minutes, repeated day and night. Males of the same population all sing variations of the same song, which changes each year. They form social groups with up to 7 members and males often fight over females. **Development** Sexually mature at 4–5 years, they live to 50 years or more. Females produce a single calf every 2–3 years, giving birth from June to October, 11–11.5 months after mating. Calves are 4–5 m long, weigh up to 2 tonnes, and are expert swimmers. They suckle for 5–7 months, drinking about 600 litres of fatty milk daily, pumped from huge mammary glands at the base of the mother's pectoral fins. Calves swim with their mother for 2–3 years. **Diet** Antarctic krill and other swimming crustaceans, fish, plankton. **Habitat** Polar to tropical coastal waters. **Status** Vulnerable. Threatened by hunting and habitat degradation.

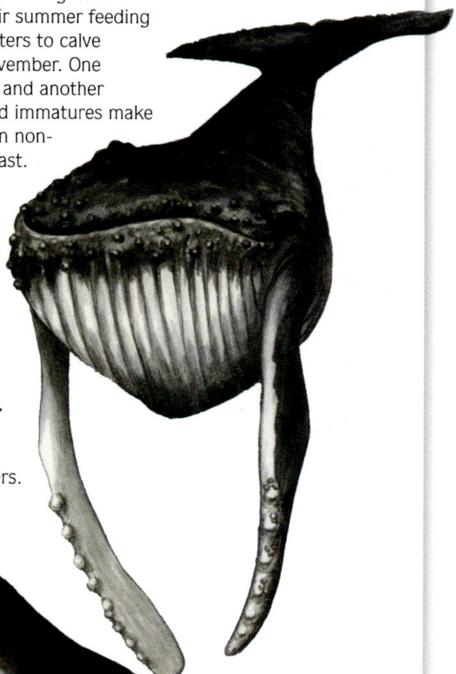

Weight **to 45 tonnes** Length **11–17.4 m**

Index

Q

R